Digital Twin Technology - Fundamentals and Applications

Edited by Orhan Korhan

Published in London, United Kingdom

Digital Twin Technology - Fundamentals and Applications
http://dx.doi.org/10.5772/intechopen.104048
Edited by Orhan Korhan

Contributors
Patrizio Armeni, Anna Gatti, Severino Meregalli, Leonardo Maria De Rossi, Irem Polat, Lorenzo Diaferia, Giacomo Visioli, Alencar Bravo, Darli Vieira, Eldrey Seolin Galindo, Urbano Chagas, Thiago Lopes da Silva, Orhan Korhan

First published in London, United Kingdom, 2023 by IntechOpen
IntechOpen is the global imprint of INTECHOPEN LIMITED, registered in England and Wales, registration number: 11086078, 5 Princes Gate Court, London, SW7 2QJ, United Kingdom

British Library Cataloguing-in-Publication Data
A catalogue record for this book is available from the British Library

Additional hard and PDF copies can be obtained from orders@intechopen.com

Digital Twin Technology - Fundamentals and Applications
Edited by Orhan Korhan
p. cm.

This title is part of the Industrial Engineering and Management Book Series, Volume 2 Topic: Production Engineering
Series Editor: Fausto Pedro Garcia Marquez
Topic Editor: Orhan Korhan

Print ISBN 978-1-83769-306-1
Online ISBN 978-1-83769-307-8
eBook (PDF) ISBN 978-1-83769-308-5
ISSN 3029-0511

For EU product safety concerns:
IN TECH d.o.o., Prolaz Marije Krucifikse Kozulić 3, 51000 Rijeka, Croatia,
info@intechopen.com or visit our website at intechopen.com.

IntechOpen Book Series

Industrial Engineering and Management

Volume 2

Aims and Scope of the Series

Industrial Engineering and Management (IEM) is a discipline that focuses on optimizing complex processes and systems within various industries. It involves the integration of engineering, business, economics, mathematics, and behavioral sciences to improve efficiency, productivity, quality, and overall performance in organizations. Key aspects of Industrial Engineering and Management include: Process Optimization; System Analysis and Design; Quality Control and Management; Supply Chain Management; Operations Management; Human Factors and Ergonomics; Project Management; Cost Analysis and Financial Management; Decision Analysis.

Overall, Industrial Engineering and Management aims to optimize resources, improve processes, enhance productivity, and ensure the effective and efficient utilization of all elements involved in the production or delivery of goods and services. It is crucial in today's competitive business environment for organizations to stay efficient and competitive.

Production Engineering and Operational Excellence are fields of study and practices that focus on optimizing and improving the manufacturing and production processes within an organization. It combines principles from engineering, management, and operational strategies to enhance productivity, efficiency, quality, safety, and sustainability in the production of goods and services.

Here are the key components of Production Engineering and Operational Excellence: Process Optimization; Operational Excellence; Manufacturing Systems Design; Quality Management; Supply Chain Optimization; Production Planning and Scheduling; Automation and Technology Integration; Health, Safety, and Environmental Management; Cost Management; Performance Measurement and Key Performance Indicators (KPIs); Continuous Improvement and Innovation.

Production Engineering and Operational Excellence are crucial for organizations aiming to stay competitive in the global market by achieving high levels of efficiency, quality, and customer satisfaction while optimizing resources and minimizing waste. It is a multidisciplinary approach that encompasses engineering principles, management strategies, and the effective use of technology to drive operational success.

Meet the Series Editor

Fausto Pedro Garcia Marquez is a Full Professor at UCLM, Spain, with accreditation since 2013. He also holds the position of Honorary Senior Research Fellow at Birmingham University, UK, and serves as a Lecturer at the Postgraduate European Institute. In addition to these roles, Fausto has experience as a Senior Manager at Accenture from 2013 to 2014. He earned his European Ph.D. with the highest distinction. Throughout his career, Fausto has received numerous awards and honors. These include the Nominate Prize (2022), Gran Maestre (2022), Grand Prize (2021), Runner Prize (2020), and Advancement Prize (2018), as well as Runner (2015), Advancement (2013), and Silver (2012) by the International Society of Management Science and Engineering Management (ICMSEM). He was also the recipient of the First International Business Ideas Competition 2017 Award. Fausto's contributions extend to academic publishing, with over 242 papers to his name. Notably, his work has been recognized in journals like "Applied Energy" (Q1, IF 9.746, Best Paper 2020) and "Renewable Energy" (Q1, IF 8.001, Best Paper 2014). His affiliations include the editorial and authorship roles in more than 50 books, with publications through respected publishers such as Elsevier, Springer, Pearson, Mc-GrawHill, IntechOpen, IGI, Marcombo, and AlfaOmega. He has authored over 100 international chapters and holds 6 patents. Fausto serves as the Editor of 5 International Journals and is a Committee Member for more than 70 International Conferences. His research portfolio encompasses being the Principal Investigator in 4 European Projects, 8 National Projects, and participating in over 150 projects involving universities and companies. His areas of expertise and research interests span Artificial Intelligence, Maintenance, Management, Renewable Energy, Transport, Advanced Analytics, and Data Science. Fausto is a recognized Expert in the European Union in AI4People (EISMD) and ESF. He also serves as the Director of www.ingeniumgroup.eu, holds the status of Senior Member at IEEE since 2021, and has been honored as an Honorary Member of the Research Council of the Indian Institute of Finance since 2021. Fausto is also the Committee Chair of The International Society for Management Science and Engineering Management (ISMSEM) since 2020.

Meet the Volume Editor

Orhan Korhan graduated with a BS from Eastern Mediterranean University (EMU), Northern Cyprus, in 2000, an MS from the University of Louisville, USA, in 2002, and a Ph.D. in Industrial Engineering from EMU in 2010. He has been working at EMU since 2009. He became an assistant professor in 2010, an associate professor in 2014, and a full professor in 2020. He is assigned to scientific committees at several international conferences and has published several books, book chapters, and papers. He established the Industry 4.0 Research Center in EMU. He is also a certified specialist in occupational safety and health. His current research interests include Industry 4.0, work-related musculoskeletal disorders, cognitive ergonomics, and facilities planning and design.

Contents

Preface

This book introduces a deep discussion of the world of digital twins, a cutting-edge innovation that is revolutionizing the digital landscape. The idea of digital twins has emerged as a beacon of innovation, providing previously unheard-of options and possibilities as we stand at the dawn of a new era where the physical and virtual worlds seamlessly merge.

In this book, we travel deep into the digital twin technology and examine the myriad ways it can be used in different fields. Digital twins have moved beyond their technical roots to become a catalyst for reinventing how we perceive, interact with, and maximize the environment around us in industries ranging from manufacturing and health care to smart cities and beyond.

We start by studying the fundamentals of digital twins, including their history, evolution, and the enabling technology for their existence. We maneuver through the complexities of sensor networks, advanced analytics, and data integration that give these virtual counterparts life.

We uncover the various uses for digital twins as we go along. Digital twin technology has a broad and ever-expanding range of applications, from improving industrial processes and forecasting equipment failures to transforming health care through patient-specific models. We carefully examine case studies, real-world applications, and success tales that demonstrate the concrete advantages gained by businesses adopting this digital frontier.

This book intends to be your guide whether you are an experienced expert looking to improve your understanding of digital twins or an inquisitive mind intrigued by the future possibilities. Let us imagine a future when the lines between the physical and digital worlds dissolve, creating a more interconnected, intelligent, and resilient environment as we make our way through the complex tapestry of digital twin technologies.

To sum up, this book sets out on an exciting investigation of the present and potential future applications merged into the fabric of digital twin technology. It promises to be both educational and motivating, providing a glimpse into a time when the physical and virtual worlds combine to create a brand-new paradigm for creativity and advancement.

Orhan Korhan
Department of Industrial Engineering,
Eastern Mediterranean University,
Famagusta, North Cyprus, Turkey

Section 1

Introduction

Chapter 1

Introductory Chapter: Digital Twin Technology

Orhan Korhan

1. Introduction

Simulation has been used as a modeling tool to build an infrastructure for monitoring the properties of the actual system by moving data from a physical system already existing in the real world to a virtual environment. Because it can make the evolution of manufacturing processes traceable, it offers benefits in terms of time, cost, and risk management. In order to organize the required preparations, the simulation's goal is to forecast probability in the virtual environment. By simulating every aspect of the physical system in the digital world, a successful simulation is feasible [1].

The idea of digital twins has evolved as a transformational paradigm that spans the divide between the physical and digital worlds in the constantly changing technological context. A virtual representation of a physical system, process, or item is referred to as a "Digital Twin," and it is made by gathering and fusing real-time data from numerous sources. This virtual version of the real counterpart acts as a potent tool for tracking, examining, and reproducing that counterpart's actions. Digital twins allow improved comprehension, predictive insights, and well-informed decision-making across a variety of sectors by offering a mirror copy of the real-world entity [2].

They include not only the geometric features of the physical object but also its functional and behavioral traits. The Digital Twin delivers a dynamic reflection of its physical counterpart by utilizing real-time data gathered from sensors, simulations, and historical records. Therefore, data analysts and IT specialists may mimic them before producing actual gadgets. They thereby impact the development of technologies like the internet of things (IoT), artificial intelligence (AI), and data analytics in addition to being employed in manufacturing [1, 3].

Real data regarding an actual thing or system is the input for a digital twin. Based on these inputs, it then provides simulations or predictions of how the real item or system would behave. It is a computer software that can simulate in its most basic form. A digital twin is first programmed, frequently by professionals in data science or applied mathematics. These professionals start by looking at the simulated versions of genuine objects or systems. The digital twin, a mathematical model that simulates the real world, may then be created using this data [4].

2. Evolution of digital twins

The concept of digital twins has its origins in NASA's use of computer models to simulate and manage space missions in the 1960s. Digital twins, on the other hand, only became well-known recently because of developments in sensor technology, data analytics, and cloud computing. The contemporary idea of digital twins was created as a result of the convergence of various technologies, which made it possible to gather, integrate, and analyze data in real time [5].

The term "digital twin" refers to a virtual duplicate or digital representation of a real object, such as a system, process, or product. Due to its potential to fundamentally alter how we build, monitor, and optimize real-world systems, this idea has attracted a great deal of interest from a wide range of businesses. In order to realize the goal of Industry 4.0, which calls for smart factories, intelligent infrastructure, and effective supply chains, a critical enabler has been identified as the digital twin idea [6].

3. The components of a digital twin

A complete digital twin comprises the following elements:

- Physical Entity: The real physical system, item, or process that the digital twin represents is referred to here. It could be a piece of equipment used in manufacturing, a structure, a vehicle, or even the complete infrastructure of a city [1, 2].

- Virtual Representation: Digital representation of the physical entity is included in the virtual model. Geometry, material characteristics, behavioral algorithms, and other pertinent characteristics are all included [6].

- Sensors and Data Sources: These are built into the physical object to gather current information about its performance, surroundings, and operating circumstances. The virtual model is regularly updated by these data sources, maintaining synchronization [4].

- Data Integration and Analytics: Using cutting-edge analytics methods like machine learning and artificial intelligence, the gathered data is combined and processed. Understanding the behavior, patterns, and potential problems of the physical thing is made possible by this study [7].

- Tools for Visualization and Simulation: Users may engage with the digital twin using a variety of visualization and simulation tools, which offer a dynamic depiction of the state and behavior of the real item [3].

- Connectivity and IoT: The digital twin depends on a smooth data transfer between the actual object and its virtual equivalent. This connectedness is made possible via the Internet of Things (IoT), which makes it possible for data to be sent, received, and evaluated [6].

4. Applications of digital twins

Applications for digital twins may be found in many different industries:

- Manufacturing: Digital twins are used to improve manufacturing processes, track the health of the equipment, and anticipate maintenance requirements, leading to greater productivity and less downtime [8, 9].

- Healthcare: By mimicking physiological processes and forecasting illness progression, digital twins in medicine help in the tailored treatment of patients [7].
- Urban Planning: Cities may use digital twins to predict traffic patterns, energy use, and infrastructure growth, assisting in environmentally friendly urban planning [4, 6].
- Aerospace: Digital twins of aircraft and spacecraft allow for real-time performance analysis, proactive maintenance, and enhanced safety [9].
- Energy: Digital twins improve the integration of renewable energy sources, estimate energy consumption, and optimize the operation of power plants [8, 9].

5. Advantages, challenges, and future directions

The potential of the Digital Twin idea to transform businesses and fundamentally alter how we interact with the real world becomes increasingly clear as it gets popularity. They have enormous promise, but there are a number of issues that need to be resolved.

The difficulties of building realistic virtual models, interoperability across diverse systems, and data security and privacy issues are a few of the difficulties that academics and practitioners are actively attempting to solve [10].

The idea of digital twins is anticipated to develop further as technology advances, including developments in augmented reality, virtual reality, and AI-driven analytics [8, 9, 10].

6. Conclusion

Digital twin technology has caused a paradigm change in how we view, engage with, and utilize the physical environment. The origins, elements, uses, and difficulties of the idea of "digital twins" are discussed in this chapter as an introduction. The technological details, case examples, and the developing landscape of digital twin technologies will be covered in greater detail in later chapters.

Author details

Orhan Korhan
Department of Industrial Engineering, Eastern Mediterranean University, Famagusta, North Cyprus, Turkey

*Address all correspondence to: orhan.korhan@emu.edu.tr

References

[1] Korhan O, Fallaha M, Murat Çınar Z, Zeeshan Q. The Impact of Industry 4.0 on Ergonomics. London, UK: IntechOpen; 2023. DOI: 10.5772/intechopen.108864

[2] Crespi N, Drobot AT, Minerva R. The digital twin: What and why? In: The Digital Twin. Cham: Springer International Publishing; 2023. pp. 3-20

[3] Soori M, Arezoo B, Dastres R. Digital twin for smart manufacturing, a review. Sustainable Manufacturing and Service Economics. 2023;**2**:100017. DOI: 10.1016/j.smse.2023.100017

[4] Lu Y, Xu X. A review of digital twin for smart manufacturing: Trends, challenges and opportunities. The International Journal of Advanced Manufacturing Technology. 2020;**108**(7-8):2267-2279

[5] Glaessgen EH, Stargel DS. The Digital Twin Paradigm for Future NASA and US Air Force Vehicles. Honolulu, Hawaii, USA: AIAA; 2012. pp. 2012-4343

[6] Grieves M. Digital twins: Virtually everything about almost anything. The Bridge. 2002;**32**(1):4-11

[7] Tao F, Cheng J, Qi Q, Zhang M, Zhang H, Sui F. Digital twin-driven product design, manufacturing and service with big data. The International Journal of Advanced Manufacturing Technology. 2018;**94**(9-12):3563-3576

[8] Tao F, Cheng Y, Zhang L, Nee AY, Srinivasan R. Digital twins and cyber-physical systems toward smart manufacturing and industry 4.0: Correlation and comparison. Journal of Manufacturing Science and Engineering. 2018;**140**(6):061004

[9] Fischer M, Shtub A. Digital twins and cyber-physical systems in smart manufacturing and industry 4.0: A comprehensive review. Journal of Manufacturing Systems. 2020;**57**:158-172

[10] Wang Z, Wan J, Zhang D, Li D, Zhang C. Towards smart factory for industry 4.0: A self-organized multi-agent system with big data based feedback and coordination. Computer Networks. 2016;**101**:158-168

Section 2

Development of Digital Twin Technology

Chapter 2

How Digital Twins Is Being Used in Industry 4.0

Thiago Lopes da Silva and Urbano Chagas

Abstract

The impact of the 4th industrial revolution, or Industry 4.0, has generated innovations that help industrial automation, promoting the digitization of activities and processes that result in increased productivity, competitiveness, improved quality of products created, and increased capacity for companies to invoke through the use of technologies such as smart cities, energy, oil and gas, Internet of things, digital and auditory manufacturing, digital twins and systems integration, among the most diverse areas inserted in Industry 4.0. This chapter will present a summary of how the most diverse industry sectors such as smart cities, oil, gas sector, energy and other areas are applying digital twins as a tool to support the digitization of companies.

Keywords: digital twins, industry 4.0, smart cities, oil and gas, energy, opportunity, applicability

1. Introduction

The concept of digital twins, originated by NASA, emerged in 2012 when it became necessary to create virtual environments that corresponded to the data of physical objects, assisting the company in decision-making [1].

Based on the article developed by Dashkina et al. [2], it is identified that Industry 4.0 is strongly linked to digital twin technology, which corresponds to the particularity of linking the behavior of physical objects such as manufacturing lines, robots, and technical installations through the use of computers and software. The use of the digital twins concept is particularly relevant when it involves mechanical parts where their properties undergo physical wear and tear over time or even cracks in metal components. It is suggested to create monitoring routines for these components with the aim of creating process improvements through continuous feedback on the state of the parts.

Canedo [3] mentions that representing real objects in digital twins adds great value compared to local optimization as is done today. Optimization and efficiency are gains to be noted when using digital twins at the system-of-systems level, and according to the article [4], the use of digital twins is a significant factor in decision-making to positively impact product construction.

According to Pethuru and Preetha [5], through real data of objects, it is possible to create simulations that assist in predicting how physical objects will be affected based on the received data. In other words, constructing monitoring of physical objects in a controlled virtual area, it helps in the prevention and control of possible failures encountered during the use of real objects [1].

The power of using digital twins in Industry 4.0 can be explored in various areas such as IoT, smart cities, energy, oil and gas, and healthcare as a decision-making tool to assist in the continuous improvement of the process of building a sustainable business model.

2. Industry 4.0 and digital twins in the sectors

In this section, we will explore how digital twins are being applied in Industry 4.0 in the energy, oil and gas, smart cities sectors, and other areas.

2.1 Energy sector

Through research conducted in article [6] by Arowoiya, it was observed that the applicability of digital twins (DT) in the energy sector can assist in energy management, usage, and simulation creation in conjunction with real-world data [7–9] to predict potential issues found in buildings.

Another application of DT in the energy sector is related to temperature control in physical environments, where a large number of variables were manually controlled using thermometers, hydrometers, and anemometers as input for manual decision-making [6]. To address this issue, Escandón et al. [10] mentions the use of a neural intelligence network in conjunction with DT.

Simultaneously, to address the monitoring issue in buildings, ref. [6] cites article [11] that uses DT in conjunction with the creation of an integrated building information modeling (BIM) system with IoT, which sends alerts to operators, thus assisting in real-time monitoring of physical objects.

The increasing interest in academia and the industry regarding the use of DT in the energy sector stems from the possibility of real-time monitoring of an electrical network with the assistance of IoT and AI.

The energy sector can explore the creation of simulation software to assess the wear, performance, and associated costs of using specific equipment for energy production, whether on a small or large scale, by utilizing a cluster of computers.

2.2 Oil and gas sector

For ref. [12], the major risk of oil and gas extraction up to the platforms is extremely complex and risky, with the possibility of setbacks that can result in financial and catastrophic losses. It is necessary for the industry to take preventive measures to mitigate the risks.

With this issue, it is important for companies in the field to invest in technology and innovation in order to expand oil and gas extraction while mitigating the inherent risks of the circumstances. In the article [13], the author states that there are thousands of sensors, complex components, and processes to be followed, and Digital Twins (DT) can be used to assist the oil and gas sector in risk mitigation through predictive analysis of the data exposed by real objects.

In ref. [12], it is mentioned that DT can help the oil and gas sector by creating virtual environments with real data that assist administrators in testing deviations, recording and analyzing data, and advancing with the security of the business cycle.

By using the virtual environment with real-time data from real objects, DT can perform routines following a predefined process in conjunction with the assistance of AI to identify potential anomalies predictively, helping the process become increasingly secure and scalable.

One way to obtain data from oil and gas platforms, according to Priyanka et al. [12], is through the use of smart IoT modules that are installed between sensors and control points using the Routing Protocol for Lossy organizations (RPL) protocol, which will help make oil and gas extraction safer through predictive information. Additionally, according to Wanasinghe et al. [14], it helps the platform operator visualize risks in a centralized visual manner, which can aid in intervening in potential issues.

In **Figure 1** [14], it is mentioned that there are several frameworks developed for DT where a significant portion converges to include three major sections. The physical section includes accessories, sensors, and actuators. The virtual section includes multi-physics spaces, model simulations that contribute to data analysis, and finally, the connection between the physical and virtual spaces, which ensures the exchange of information between the two.

Another approach, according to Wanasinghe et al. [14], would be to create a DT framework using a five-component model, as shown in **Figure 2**. The physical environment contains all the physical accessories, sensors, and actuators. The virtual space contains a mirror of the physical environment for high fidelity in creating simulations. The service system contains another enterprise application responsible for service visualizations, service quality, diagnostic services, model calibration, algorithm services, and other services. The DT data fusion acts as a bridge between the physical environment, virtual model, and service system.

According to Wanasinghe et al. [14], the use of machine learning, deep learning, and artificial intelligence with intelligent mathematical algorithms will be important tools in the predictive evaluation of possible risks associated with the components used on platforms, thereby avoiding potential failures and accidents.

The oil and gas sector can explore simulation programs that can be created using real data and characteristics of the components, especially the variables of the environment that are part of the oil extraction process, to assist in the predictive identification of wear and tear through the use of mathematical models that help the company mitigate risks, costs, environmental problems, and assist in the scheduled replacement of components. In this context, it is relevant for the company to create scalable solutions in machine clusters due to the possibility of a high level of data processing that may be generated during the prediction process.

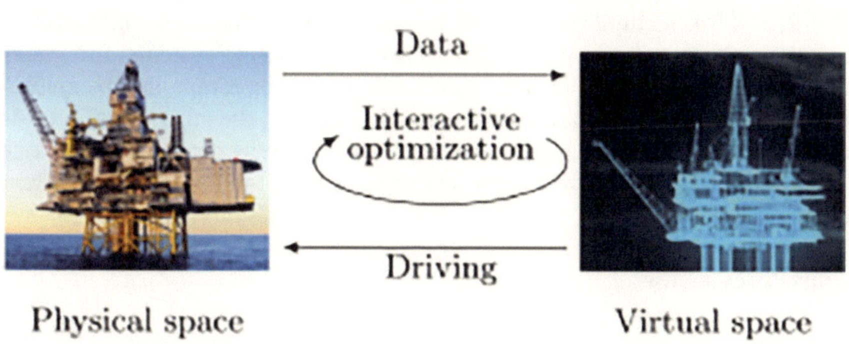

Figure 1.
Digital twin framework with three components (physical space, virtual space, and connection between them) [14].

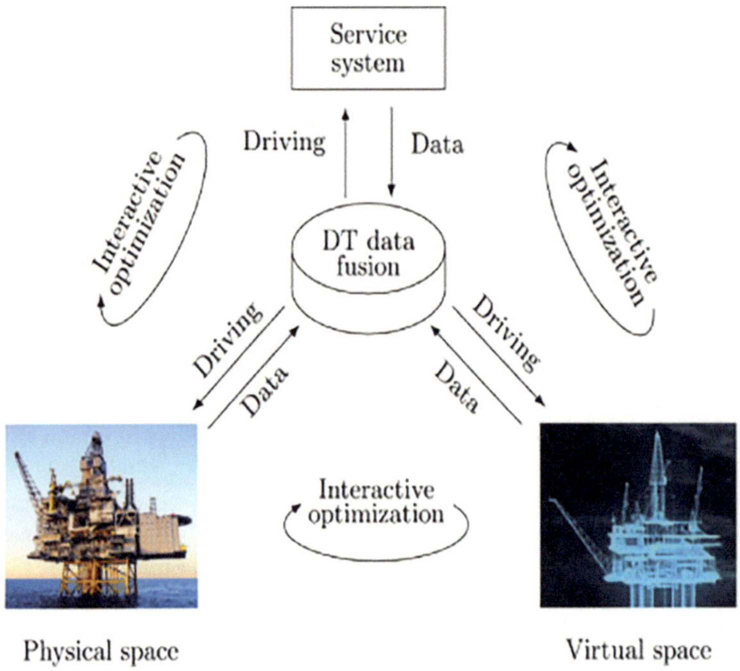

Figure 2.
Digital twins framework with five components (physical space, virtual space, connection between them, data and service) [14].

Wanasinghe et al. [14] mentions that simulations using DT will assist engineers in building new platforms or modifying their oil and gas extraction structures, reducing risks, as this step can be repeated whenever necessary to bring greater reliability, performance, and reduction of unnecessary costs before executing the operation.

Wanasinghe et al. [14] observed that DT can help engineers and individuals involved in problem-solving in a faster and cooperative way by obtaining real-time information from the platforms, enabling the creation of virtual rooms with the possibility of simulations to improve performance in critical decision-making for any occurring problems.

Another factor where DT can help in the oil and gas sector, according to Wanasinghe et al. [14], is the creation of training centers in conjunction with the use of VR, AR, and MR technologies to train new employees who can navigate within oil and gas centers with ease of operating equipment, inspecting systems, and interacting with ongoing operations. This is a significant factor for companies to explore this possibility since, according to research in articles [15–16], the oil and gas sector is facing a challenge called the "big-crew change," where in the near future, more than 50% of experienced workers will retire, causing the sector to lose skills and talents in the industry.

2.3 Smart cities sector

According to Zhuang et al. [17], the characteristics of Digital Twins (DT), such as the integration of various types of data from physical objects, involvement throughout the lifecycle of physical objects, co-evolution with them, and continuously accumulating relevant knowledge, can help the government [18] create more predictive and comprehensive prediction and indicators in a smart cities ecosystem.

In the field of smart cities, DT, as mentioned in ref. [17], can be used to build physical maps of the real city in a virtual area that receives real-time events from the mapped objects. This allows for transferring, modifying, deleting, and performing operations in a city area through the created 3D models, while checking for possible problems that may occur through these operations.

Mohammadi and Taylor [19] mentions a project being developed using DT concepts in smart cities called the Digital Twin City of Atlanta, which utilizes a VR platform developed with Unity, a cross-platform framework designed for video games. This environment helps discover interactions and interoperability of its human infrastructure systems.

According to Ivanov et al. [20], DT can bring opportunities for improvements in smart cities through observability of resident traffic flow, private business traffic, public transportation, and real-time information from private and public intelligent sensors that help monitor and analyze temperature and humidity as shown in **Figure 3**.

According to Ivanov et al. [20], through the collected data in a smart city, the following opportunities for improvements can be generated along with the use of IoT:

- Creating a DT of the public transportation network to monitor and predict possible availability and efficiency situations of transportation.

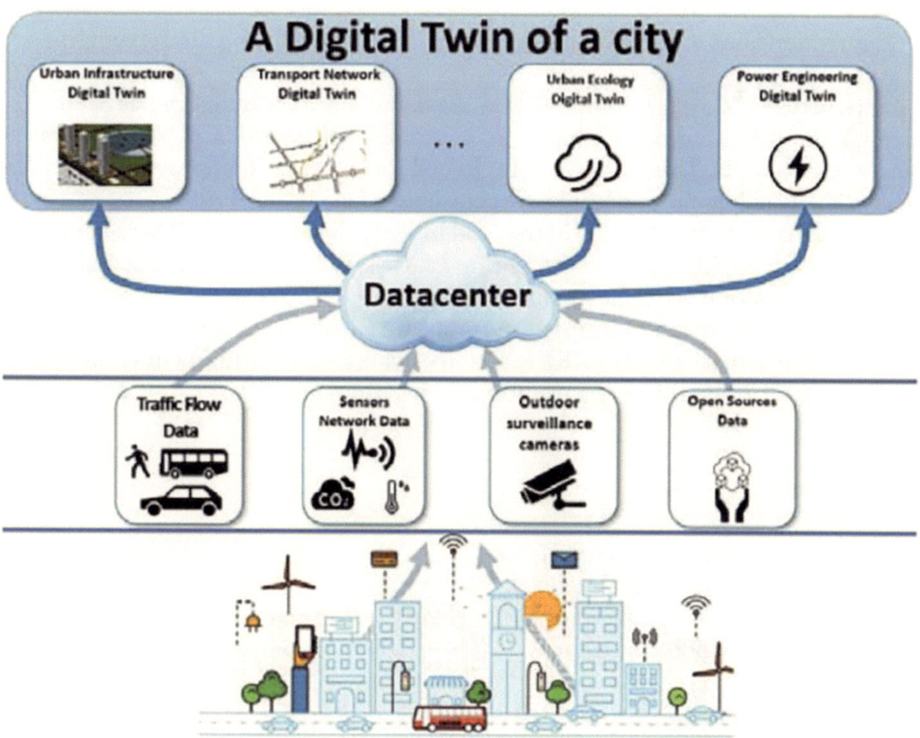

Figure 3.
An example of the interface's appearance of the use of digital twins in smart cities [20].

- Creating a DT of the city's ecological system to promote monitoring and predict changes in the environment, including changes in soil, water, air quality, etc.
- Improving the efficiency of city design solutions evaluation.
- Identifying potential sources of risks.
- Controlling pollution through continuous environmental analysis.
- Energy rationing through on-demand usage of city energy.
- Improving efficiency and reducing costs through on-demand waste removal instead of scheduled waste collection.

To analyze the large amount of data received by sensors in a smart city within a DT, it is necessary, as stated in ref. [21], to include statistics, data analysis intelligence, and a computational model.

Ivanov et al. [20] states the need to create a large warehouse that can handle a significant amount of data, sufficient bandwidth to collect and analyze the data received, and computational power to support the high degree of processing that can be done through techniques such as machine learning.

2.4 Other areas

This section aims to show how some other areas are exploring digital twins in various sectors.

In the article [22], it is mentioned that the manufacturing industry has been using DT through monitoring, simulations, and remote control of physical assets using virtual objects. This, in turn, helps in understanding and improving customer satisfaction by enhancing existing products, operations, and services.

Also, in the same article [22], in the field of agriculture, it is possible to use DT by creating virtual environments representing a farm with the goal of increasing productivity and production efficiency while reducing energy and costs.

Regarding education and training, Attaran and Celik [22] explains that the use of DT through Virtual Reality (VR) has been assisting in the training of doctors by complementing and refining the traditional educational model.

3. Discussion

In this chapter, which focused on the study in the areas of oil and gas, energy, smart cities, and other general fields, the observation was made on how the industry has been using DT (Data Technology) to enhance process quality through the following main characteristics.

3.1 Usage of real data in virtual environments

The use of real data is connected and directly utilized in a virtual environment, thereby assisting in the creation of simulations that aid in better decision-making.

3.2 Virtual reality (VR)

The use of virtual reality has aided in the development of more effective training and improved interactions with real objects in a virtual world.

3.3 Monitoring

Real-time monitoring of physical objects in a virtual environment will help in predicting potential issues through the use of IA.

4. Conclusion

In Industry 4.0, one of the pillars being explored in the industry is Digital Twins. This work aimed to demonstrate how Digital Twins are being used in the sectors of smart cities, energy, oil and gas, and other areas, and how they have helped in reducing risks, and costs, improving processes, and enabling real-time monitoring of physical objects through the use of the Internet of Things and artificial intelligence.

Thanks

I would like to thank my family, Urbano Chagas and CESAR for the opportunity to share about the topic discussed.

Abbreviations

DT	digital twins
BIM	building information modeling
IoT	Internet of Things
VR	virtual reality
AR	Augmented reality

Author details

Thiago Lopes da Silva* and Urbano Chagas
Centro de Estudos e Sistemas Avançados do Recife (CESAR), Recife, Brazil

*Address all correspondence to: tls@cesar.org.br

References

[1] Lin YW, Tang TLE, Spanos CJ. Hybrid approach for digital twins in the built environment. In: Proceedings of the Twelfth ACM International Conference on Future Energy Systems, e-Energy 21. New York, NY, USA: Association for Computing Machinery; 2021. pp. 450-457. ISBN 9781450383332. DOI: 10.1145/3447555.3466585

[2] Dashkina A, Khalyapina L, Kobicheva A, Lazovskaya T, Malykhina G, Tarkhov D. Neural Network Modeling as a Method for Creating Digital Twins: From Industry 4.0 to Industry 4.1. New York, NY, USA: Association for Computing Machinery; 2021. ISSN 9781450388313. DOI: 10.1145/3444465.3444535

[3] Canedo A. Industrial IoT lifecycle via digital twins. In: Proceedings of the Eleventh IEEE/ACM/IFIP International Conference on Hardware/Software Codesign and System Synthesis, CODES 16. New York, NY, USA: Association for Computing Machinery; 2016. ISBN 9781450344838. DOI: 10.1145/2968456.2974007

[4] Lim KYH, Zheng P, Chen C. A state-of-the-art survey of digital twin: Techniques, engineering product lifecycle management and business innovation perspectives. Journal of Intelligent Manufacturing. 2020;**31**:1313-1337. DOI: 10.1007/s10845-019-01512-w

[5] Pethuru R, Preetha E. Digital Twin: The Industry Use Cases Pethuru Raja. Tiruchirappalli, India: Chellammal Surianarayanana, a Reliance Jio Infocomm Ltd. (RJIL), Bangalore, India Bharathidasan University Constituent Arts & Science College; 2020

[6] Arowoiya VA, Moehler RC, Fang Y. Digital twin technology for thermal comfort and energy efficiency in buildings: A state-of-the-art and future directions. In: Energy and Built Environment. 2023. DOI: 10.1016/j.enbenv.2023.05.004. ISSN 2666-1233. Available from: https://www.sciencedirect.com/science/article/pii/S2666123323000314

[7] Jafari MA, Zaidan E, Ghofrani A, Mahani K, Farzan F. Improving building energy footprint and asset performance using digital twin technology. In: IFAC-PapersOnLine. 4th IFAC Workshop on Advanced Maintenance Engineering, Services and Technologies - AMEST 2020. Vol. 53, no. 3. 2020. pp. 386-391. DOI: 10.1016/j.ifacol.2020.11.062. ISSN 2405-8963. Available from: https://www.sciencedirect.com/science/article/pii/S2405896320302123

[8] Clausen A et al. A digital twin framework for improving energy efficiency and occupant comfort in public and commercial buildings. Energy Informatics. 2021;**4**(2):40

[9] Francisco A et al. Occupant perceptions of building information model-based energy visualizations in eco-feedback systems. Applied Energy. 2018;**221**:220-228

[10] Escandón R et al. Thermal comfort prediction in a building category: Artificial neural network generation from calibrated models for a social housing stock in southern, Europe. Applied Thermal Engineering. 2019;**150**:492-505

[11] Valinejadshoubi M et al. Development of an IoT and BIM-based automated alert system for thermal comfort monitoring in buildings. Sustainable Cities and Society. 2021;**66**:102602

[12] Priyanka EB, Thangavel S, Gao XZ, Sivakumar NS. Digital twin for oil pipeline risk estimation using prognostic and machine learning techniques. Journal of Industrial Information Integration. 2022;**26**:100272. ISSN 2452-414X. DOI: 10.1016/j.jii.2021.100272. Available from: https://www.sciencedirect.com/science/article/pii/S2452414X21000704

[13] Strasser T, Andrén F, Lehfuss F, Stifter M, Palensky P. Online reconfigurable control software for IEDs. IEEE Transactions on Industrial Informatics. 2017;**9**(3):1455-1465

[14] Wanasinghe TR, Wroblewski L, Petersen BK, Gosine RG, James LA, de Silva O, et al. Digital twin for the oil and gas industry: Overview, research trends, opportunities, and challenges. In: IEEE Access. Vol. 8. 2020. pp. 104175-104197. DOI: 10.1109/ACCESS.2020.2998723

[15] PeeK P, Fenard J, Gantes P, Theiler C. Skills Shortages in the Global Oil and Gas Industry—How to Close the Gap (Part I). Geneva, Switzerland, Tech. Rep.,: CRES; 2008

[16] Parshall J. After years, 'big crew change' has passed, but learning, train-ing challenges remain. Journal of Petroleum Technology. 2017;**69**(7):38-40

[17] Zhuang C, Liu J, Xiong H. Digital twin-based smart production management and control framework for the complex product assembly shop-floor. International Journal of Advanced Manufacturing Technology. 2018;**96**(1e4):1149e1163. DOI: 10.1007/s00170-018-1617-6

[18] Deng T, Zhang K, Shen ZJ(M). A systematic review of a digital twin city: A new pattern of urban governance toward smart cities. Journal of Management Science and Engineering. 2021;**6**(2):125-134. DOI: 10.1016/j.jmse.2021.03.003

[19] Mohammadi N, Taylor JE. Smart city digital twins. In: 2017 IEEE Symposium Series on Computational Intelligence (SSCI). 2017. pp. 1-5. DOI: 10.1109/SSCI.2017.8285439

[20] Ivanov S, Nikolskaya K, Radchenko G, Sokolinsky L, Zymbler M. Digital twin of city: Concept overview. In: 2020 Global Smart Industry Conference (GloSIC). 2020. pp. 178-186. DOI: 10.1109/GloSIC50886.2020.9267879

[21] Korambath P, Wang J, Kumar A, Davis J. A smart manufacturing use case: Furnace temperature balancing in steam methane reforming process via Kepler workflows. Procedia Computer Science. 2016;**80**:680-689. DOI: 10.1016/j.procs.2016.05.357

[22] Attaran M, Celik BG. Digital twin: Benefits, use cases, challenges, and opportunities. Decision Analytics Journal. 2023;**6**:100165. ISSN 2772-6622. DOI: 10.1016/j.dajour.2023.100165. Available from: https://www.sciencedirect.com/science/article/pii/S277266222300005X

Chapter 3

Perspective Chapter: Digital Twin Technology as a Tool to Enhance the Performance of Agile Project Management

Alencar Bravo and Darli Vieira

Abstract

In this chapter, we examine the intersection of two paradigm-shifting ideas that are reshaping the contemporary landscape of business: agile project management and digital twin technology. We initially review the basis of agile project management, with a focus on the approach that is iterative, adaptive, and customer-centric. On this basis, we examine the role of digital twins in facilitating effective communication and coordination within cross-functional agile teams. The synergy between digital twins and agile project management has been explored, with a focus on how better decision-making, risk management, and deliverables can be facilitated within complex physical product development projects. Through the integration of digital twins into agile project management practices, organizations can achieve enhanced visibility, collaboration, and efficiency throughout the project lifecycle. In conclusion, we determined that the digital twin serves as an indispensable instrument in complex agile projects, significantly augmenting their efficacy in numerous aspects.

Keywords: agile, project, management, digital twins, collaboration, customer

1. Introduction

Within the realm of management sciences, it is of paramount importance a comprehensive understanding of the substantial distinctions that exist between projects and operations [1]. On the one hand, operational activities refer to the ongoing and repetitive tasks that are essential to sustain a business [2]. On the other hand, projects are temporary endeavors that are undertaken with the aim of achieving specific objectives [3]. As an illustration, the duty of overseeing a manufacturing facility can be characterized as an intricate operation, with the primary objective of ensuring a steady output of goods and services that meet the highest standards of excellence. In contrast, the endeavor of building a new manufacturing facility would entail a set of well-defined objectives, such as accomplishing the construction phase within the stipulated time frame and with the allocated financial resources. The significance of this differentiation lies in the fact that each of these two entities demands distinct management methodologies, instruments, and practices.

The practice of project management entails the systematic planning, meticulous organization, and efficient allocation of resources toward the attainment of predetermined objectives within well-defined limitations [4]. Project management involves a comprehensive spectrum of tasks, spanning from the conception and planning phase to the implementation, monitoring, and conclusion. The efficacy of project management is contingent upon its capacity to execute projects punctually, within financial constraints, and with the intended level of excellence [5]. The classical methodology for project management is characterized by a predictive orientation, whereby the project plan is formulated in advance and adhered to throughout the entire project duration. This particular methodology presupposes that the prerequisites of the project are comprehensively comprehended and can be precisely delineated prior to project execution. Furthermore, it assumes that the project team possesses the essential expertise and resources to execute the project effectively. Nonetheless, it is imperative to note that this particular methodology may not be appropriate for endeavors that are complex or ambiguous or that necessitate a substantial level of inventiveness [6].

The agile project management (APM) approach has surfaced as a means of managing projects in a more flexible and adaptive manner and has garnered significant traction across diverse industries [7, 8]. In the realm of project management, the concept of agility pertains to the capacity to react promptly and adeptly to alterations in project specifications, stakeholder demands, and market fluctuations [9]. APM is an approach that prioritizes iterative and adaptable processes, with a strong emphasis on collaboration and continuous improvement. This approach is designed to enable teams to respond quickly and effectively to changing circumstances and to maintain a focus on delivering high-quality results. By fostering a culture of flexibility and open communication, APM can help organizations achieve their goals more efficiently and effectively [10].

In contrast to classical project management methodologies that adhere to a linear and sequential approach, APM fosters adaptability and promotes regular feedback and iteration. As APM continues to gain ground across various industries, another technology that is becoming popular is the digital twin (DT), a computer-generated replica of a tangible entity, operation, or framework. This cutting-edge technology enables the creation of a virtual model that mirrors the real-world counterpart, allowing for the simulation of various scenarios and the analysis of data in a controlled environment. In contrast to conventional computer-aided design, this innovative approach integrates real-time data, simulation, and analytics to generate a comprehensive model that emulates the behavior and attributes of the corresponding physical entity [11].

The concept of DT has garnered significant attention in recent times and has emerged as a transformative trend for various industries, much like APM did in the past. DTs serve as dynamic reflections of physical systems, facilitating ongoing monitoring, simulation, and analysis. The information gathered from the physical system is harmoniously integrated with the DT, furnishing invaluable discernments for decision-making and optimization [12]. DT technology offers a multitude of benefits for diverse industries, enabling them to engage in virtual prototyping and predictive maintenance and optimize performance. As the potential of DT technology has become increasingly apparent, organizations have begun to incorporate it into their overall project management strategies [13]. Through the development of a virtual representation of the project's end product, teams can acquire a more comprehensive understanding of the system's behavior, proactively anticipate and resolve potential issues, and optimize the overall outcomes of the project [11, 14]. Projects are inherently complex and pose numerous challenges that can be effectively addressed through the use of DT technology. This innovative approach can significantly enhance decision-making processes, facilitate the seamless transmission of critical information, and improve risk mitigation strategies. By leveraging the power of DT technology, project stakeholders can achieve greater efficiency, accuracy, and overall success in their endeavors.

Among the plethora of proposed APM methodologies, Scrum has emerged as the preferred approach for most cases and has gained widespread adoption because of its effectiveness in managing complex and dynamic projects [15]. Scrum is a highly effective iterative framework that is designed to deliver incremental value through a series of short development cycles, which are commonly referred to as sprints. This approach is widely recognized for its ability to enhance productivity, improve collaboration, and foster a culture of continuous improvement [16]. By leveraging the Scrum methodology, organizations can achieve greater agility and responsiveness and ensure that their products and services are delivered to market in a timely and efficient manner.

In this chapter, we will delve into the intricacies of DT technology within the realm of APM, with a particular emphasis on Scrum projects. Readers will acquire invaluable insights from this chapter on how to leverage the potential of DT technology effectively in the context of APM.

We have structured this chapter in the following manner: Section 2 provides an in-depth exploration of the fundamental principles of APM, including its unique characteristics. In Section 3, the focus shifts to the practical aspects of project and product development, with a particular emphasis on the distinctions between software and hardware development. This section also addresses the challenges inherent in implementing APM in software development. Finally, in Section 4, we present DT technology as a viable solution to many of the challenges encountered in physical product development within the APM framework. This section provides a comprehensive discussion of how the DT can be successfully implemented as well as its numerous advantages, making it an indispensable tool for physical product projects within the APM paradigm.

2. Agile project management

The genesis of the APM can be traced back to the 1990s, a time when software development teams were searching for alternative project management approaches to the conventional ones [17]. Numerous industries expressed their displeasure with the inflexible and step-by-step approach of the traditional waterfall methodology [18]. It is imperative for industries to delve into innovative approaches to address the ever-changing demands of the market, stakeholders, and customer requirements [19]. The traditional approach of defining all requirements upfront and following a linear project progression may not be suitable in such unpredictable market conditions [7, 20]. The APM movement gained momentum in the midst of the ever-increasing dynamism of the market, and it was formalized and characterized by a set of principles and values outlined in the Agile Manifesto, which a group of software developers and thought leaders developed in 2001 [21, 22]. At the core of the manifesto lies the notion that APM has to prioritize the significance of individuals and their interactions, the perpetual delivery of functional software, active customer collaboration, and the ability to adapt to change.

APM methodologies are commonly recognized for their emphasis on flexibility and adaptability [23]. Regardless of the particular methodology that an individual selects, it is imperative to acknowledge that the pursuit of ongoing enhancement represents yet another pivotal facet of APM [24, 25]. It is customary for agile teams to regularly engage in introspection regarding their processes, actively solicit feedback, and pinpoint opportunities for enhancement. APM entails the cultivation of a culture that prioritizes continuous learning [26]. This approach enables teams to fine-tune their practices, elevate their performance, and ultimately achieve superior outcomes over time. Moreover, one could argue that APM is better suited for technological and complex projects because of its inherent recognition of the potential for project requirements to evolve [27]. This approach embraces change by prioritizing iterative and incremental development, allowing for greater flexibility and adaptability throughout the project life cycle. In the context of a Scrum project, the classical approach of defining all project requirements upfront is replaced by a more dynamic and collaborative process. Here, small and multidisciplinary teams operate in short cycles with the aim of delivering a potentially shippable product increment at the end of each cycle. The use of an iterative approach facilitates teams to effectively integrate feedback, modify priorities, and flexibly adapt their plans as they advance through the project [28].

Since its inception APM has evolved, resulting in the development of numerous methodologies and frameworks, all of which adhere to the fundamental principles and values outlined in the Agile Manifesto. However, they diverge in terms of their prescribed practices and tools, which are tailored to suit the unique requirements of individual project contexts. Thus, one can posit that APM comprises a diverse array of methodologies, each proffering a distinctive approach to providing projects in an iterative and collaborative fashion. As we stated earlier, Scrum stands as the most extensively implemented agile methodology. The State of Agile Report 2022 [29] showed that Scrum continues to be the most widely employed agile framework, with an impressive adoption rate of 87%. Scrum fundamentally places a strong emphasis on the formation of self-organizing teams, the execution of short iterative sprints, and the regular engagement of stakeholders in the collaborative process [30, 31]. The following sections will provide additional information regarding the Scrum framework.

Over time, APM has transcended its origins in software development and has been embraced by a diverse array of industries, including but not limited to manufacturing, health care, and marketing [7, 32, 33]. The principles and practices of the agile methodology have demonstrated their efficacy in tackling the intricacies and dynamism of complex projects [34]. Even companies that are not fully embracing the agile methodology are increasingly seeking to integrate elements of APM into their projects [35]. As an illustration, it is plausible to alter the framework and internal regulations of a conventional project to accentuate collaboration and stakeholder engagement, which is a pivotal facet of APM. Adopting a customer-centric approach empowers teams to provide products and solutions that are in sync with their customer's expectations, thereby resulting in heightened satisfaction and value [36].

3. Agile software and physical product development

In this particular section, we delve further into the application of Scrum. It is noteworthy that statistics have indicated Scrum to be at the forefront of methodologies in terms of use. The Scrum methodology is extensively employed in the development of software as well as physical products, such as hardware. Although the fundamental tenets of Scrum remain steadfast across various domains, there exist certain modifications that are particular to the realms of software and hardware development.

3.1 Scrum for software project management

Scrum provides a structured approach to overseeing the iterative and incremental advancement of software products within software development projects [37]. The approach emphasizes fostering a strong sense of collaboration, promoting transparency, and embracing adaptability as key tenets of its philosophy [38]. The Scrum framework comprises three critical roles—namely, the product owner, the development team, and the Scrum master. These roles are essential in ensuring the successful implementation of the Scrum methodology. The product owner serves as the stakeholders' representative and assumes the role of intermediary between the development team and the client. The product owner bears the responsibility of delineating and assigning priority to the product backlog, a meticulously arranged index of characteristics, improvements, and bug resolutions. The role of the product owner is to ensure that the team is focused on delivering the most valuable items that are in alignment with the overarching goals of the project. For this, it is necessary to have a keen understanding of the project's objectives and a deep appreciation for the stakeholders' needs. By prioritizing the work that needs to be done and providing clear guidance to the team, the product owner plays a critical role in ensuring that the project is successful [39].

The development team comprises a group of cross-functional individuals who deliver the product increment [40]. The composition of this team generally encompasses a diverse range of professionals, such as developers, testers, designers, and other pertinent roles. The team collaboratively addresses the chosen items from the product backlog within sprints. These sprints are brief periods of work, typically spanning two to 4 weeks. The team members engage in a process of self-organization whereby they collaboratively plan their work and make commitments for each sprint.

The role of the Scrum master entails the facilitation of the Scrum process, the removal of impediments, and ensuring the team's adherence to Scrum principles and practices. As a coach, the Scrum master adeptly guides the team toward its goals while simultaneously cultivating a highly productive and collaborative work environment [41]. The role of the Scrum master is to aid in the facilitation of various meetings, including but not limited to the daily stand-up, sprint planning, sprint review, and sprint retrospective.

The Scrum methodology employs a range of artifacts to facilitate efficient project management within the realm of software development. The product backlog acts as the only reliable source of information regarding the scope of the project and directs the work that the team does throughout the duration of the project. The sprint backlog is a crucial artifact in Scrum methodology because it comprises the items that the development team has committed to accomplishing during the sprint. These items are carefully selected from the product backlog, ensuring that they align with the team's capacity and the sprint goal. By maintaining a well-defined sprint backlog, the team can effectively plan and execute its work, ultimately delivering a high-quality product increment at the end of the sprint [42]. The sprint backlog outlines the activities that must be completed to deliver the product increment successfully, and its creation is a team effort. In the context of agile software development, the increment refers to the cumulative sum of all product backlog items that have been successfully completed in a sprint. This term is of utmost importance in the agile methodology because it represents the tangible progress made by the development team toward the ultimate goal of delivering a fully functional product. The increment denotes a product that holds the potential to be shipped and is capable of delivering significant value to the stakeholders involved [43]. Furthermore, the burndown chart serves as a visual aid that illustrates the amount of work that remains in the sprint backlog as time progresses. This tool is instrumental in enabling the team members to monitor their progress and make necessary adjustments to their efforts. An example of this type of chart can be seen in **Figure 1**.

In summary, we can state that Scrum is capable of providing a structured and collaborative approach to project management. This method enables teams to deliver high-quality software incrementally. Although Scrum is commonly recognized for its efficacy in software development, it can also be adeptly customized to facilitate the development of physical items, as we will elaborate in the following section.

3.2 Agile physical product development

The Scrum methodology has gained widespread adoption in the realm of software development [44]. However, when it comes to the development of novel physical products, certain modifications must be made to account for the distinctive challenges inherent in such an undertaking. When examining this issue, it is crucial to consider the significance of creating prototypes to facilitate the advancement of the development of tangible products [45]. Incorporating prototyping as a fundamental element of the iterative process is a crucial adaptation that must be made when implementing Scrum in the realm of physical product development. In contrast to software, the development of physical products necessitates the contemplation of a multitude of tangible elements, including but not limited to materials, form, ergonomics, and manufacturability [46]. The process of prototyping enables teams to delve into these various aspects at an early stage, thereby facilitating the identification of design flaws, usability concerns, and potential enhancements [47].

The integration of prototyping into Scrum commences with the product backlog. It is imperative the backlog take into account the inclusion of prototyping activities as

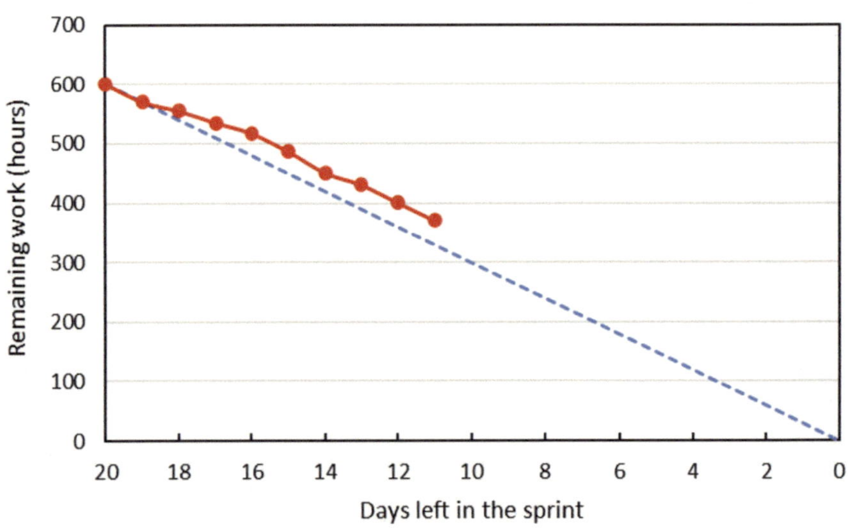

Figure 1.
Example of a burndown chart used to measure the progress of work in a sprint.

indispensable constituents of every sprint. It is imperative to allocate sufficient time and resources to the prototyping phase within the development cycle to guarantee its efficacy. The product owner, in close collaboration with the team, meticulously defines user stories and tasks that are pertinent to the prototyping activities. Throughout every sprint, the development team diligently engages in the process of constructing prototypes that are in accordance with the prioritized features and requirements. As was already seen, the team comprises individuals who possess a wealth of knowledge and experience in various fields, such as industrial design, engineering, manufacturing, and other pertinent disciplines. Together, the team works in unison to produce prototypes that effectively encapsulate the proposed product's fundamental qualities.

Thus, it is apparent that the integration of prototyping with Scrum methodology can yield numerous advantages in the realm of physical product development. To begin with, it enables teams to authenticate design concepts and evaluate functionality [48]. Through the development of preliminary models, teams have the opportunity to obtain valuable input from stakeholders, users, and prospective clients. This feedback empowers them to make well-informed judgments and refine the design and functionality of the product through iterative processes. The use of an iterative feedback loop serves as a valuable tool for teams to enhance the quality of their prototypes and ensure that the product is in sync with the requirements of the end users and the market. In addition, the process of prototyping serves as a valuable tool for assessing the potential manufacturability and production feasibility of a given physical product [49]. The development of prototypes that bear a strong resemblance to the ultimate product enables teams to evaluate a range of factors, including material compatibility, assembly processes, and cost considerations [50]. The prompt detection of manufacturing limitations or obstacles confers upon teams the ability to effectuate requisite design adjustments, thereby guaranteeing a seamless progression from prototype to production.

Therefore, prototyping combined with the Scrum framework presents a multitude of opportunities for the exhibition of prototypes and the acquisition of feedback. The sprint review is an invaluable opportunity to showcase the latest prototype to stakeholders, affording them the chance to experience the product's form and functionality directly. The feedback that is received plays a significant role in shaping the product's subsequent iterations. It serves as a driving force for improvements and helps to align the product with the desired outcomes. Moreover, the retrospective meeting within the Scrum framework offers a valuable opportunity for the team to engage in introspection regarding its prototyping practices. This space allows for the identification of areas that require improvement as well as the sharing of insights that have been gathered throughout the project. The process of iterative learning is instrumental in enabling teams to fine-tune their prototyping methodology, thereby enhancing its efficiency, efficacy, and alignment with project objectives [51].

In essence, the use of Scrum methodology for the purpose of developing novel physical products yields significant advantages, particularly when emphasis is placed on the prototyping phase. The integration of prototyping within the Scrum framework presents teams with the opportunity to validate design concepts, collect valuable feedback, and refine the product through an iterative process. Incorporating prototyping activities into the product backlog, allocating ample time and resources for prototyping iterations within each sprint, and using the sprint review and retrospective meetings to validate products and processes can enable teams to leverage all the potential of prototyping effectively. The result can be the development of superior physical products that respond to user requirements, align with market trends, and foster innovation. The employment of this tool enables teams to generate concrete depictions of their concepts and blueprints, thereby furnishing a medium for scrutinizing, assessing, and enhancing the product during the project evolution. As we delve deeper into the subject matter, it becomes apparent that the integration of DT technology into the Scrum framework can yield more significant advantages for teams, thereby increasing even more their likelihood of success.

3.3 Challenges with complex physical product development

As evidenced in the preceding section, prototypes function as concrete manifestations of the product, affording teams the opportunity to authenticate and enhance design concepts, assess functional components, and amass invaluable user input. The process of prototyping assumes a pivotal role in the development of physical products within the Scrum methodology. The use of this tool allows agile teams to expeditiously recognize and tackle potential design flaws or usability issues at the outset of the development process, thereby reducing the need for future expensive revisions [52]. However, the integration of products that distinct Scrum teams develop can pose several challenges, despite the employment of diverse prototypes. These challenges primarily pertain to interfaces and information communication. As each team concentrates on the development of specific components or modules, it is imperative to pay meticulous attention to interface design and effective communication practices to ensure seamless integration [53]. In this section, we will delve into several pivotal obstacles that pertain to interfaces and information communication.

Medium to complex projects often face multiple risks and potential issues arising from the various interface compatibilities [54, 55]. Interfaces serve as the crucial junctions that link distinct components or modules that separate teams have developed. It is of utmost importance to ensure compatibility across interfaces to facilitate a seamless integration process toward the final and comprehensive product that meets the required standards of quality. In the realm of complex projects, a multitude of challenges may arise when teams interpret interface requirements in varying ways or when modifications to one component have a consequential impact on the functionality of other components that depend on it. Effective communication and seamless collaboration among teams are crucial for establishing and sustaining compatible interfaces throughout the development process of complex projects [56].

In complex projects, interface specifications that are ambiguous or incomplete can give rise to a multitude of issues. Inconsistencies and delays may arise when multiple teams hold divergent interpretations of interface requirements. To ensure the success of complex projects, normally a significant amount of time should be dedicated to the initial phase of defining thorough and unequivocal interface specifications [57]. This approach serves to mitigate the likelihood of any potential issues arising throughout the project's lifespan. In the context of APM, it is worth noting that the aforementioned issue can be somewhat alleviated. This is because of the fact that frequent communication and clarification between teams can facilitate a collective comprehension of interface requirements, thereby circumventing any potential misalignment [58]. In complex projects, it is imperative for teams to effectively communicate updates, progress, and potential issues pertaining to their respective components. The occurrence of delays or miscommunication in the dissemination of information can impede the integration process, ultimately resulting in misunderstandings or conflicts [59, 60]. The establishment of effective communication channels, such as scheduled meetings or digital collaboration platforms, can greatly facilitate the timely and transparent exchange of information.

Nevertheless, in the case of a complex project, different Scrum teams may run different sprint cycles or have divergent development timelines. The integration of their products can pose certain challenges. The occurrence of misalignment in timelines can potentially result in undesirable delays or compel teams to operate on incomplete components, thereby adversely affecting the overall integration process [61]. One potential solution to address these challenges and facilitate a seamless integration process is to coordinate and synchronize sprint cycles or to establish clear dependencies and sequencing of work between teams. By doing so, teams can work together more effectively and efficiently, ultimately leading to a more successful outcome. In addition, it is important to note that as the development process advances, modifications made to a particular component may potentially trigger a series of consequential impacts on other components that rely on it. The task of coordinating and managing changes and dependencies can prove to be quite challenging, particularly when multiple Scrum teams are involved [62]. Well-defined change management procedures, efficient communication channels, and a cooperative strategy for handling interdependencies can significantly reduce risks and guarantee a smooth integration of modifications into the overall product [63, 64].

In summary, it is imperative to acknowledge that the seamless integration of products from diverse Scrum teams necessitates efficient knowledge dissemination and collaboration. Each team possesses a wealth of valuable insights and expertise that are directly related to their respective components. Nonetheless, the inadequate dissemination and use of this valuable knowledge has the potential to impede the straightforward assimilation process. Facilitating cross-team collaboration, promoting knowledge transfer sessions, and fostering a culture of open communication can effectively encourage the exchange of ideas, insights, and best practices. This way of working, in turn, can significantly enhance the integration of different products. As we will see in the next section, it is imperative to acknowledge that in the context of physical product development, the DT can take the form of an indispensable tool for facilitating seamless and effective knowledge transfer among teams.

4. Agile project management and digital twin: a powerful combination for complex projects

As evidenced in the preceding section, the integration of products that various Scrum teams have developed can present certain challenges, particularly regarding interfaces and the communication of information. To achieve successful integration, it is essential to address interface compatibility, ensure clear and comprehensive interface specifications, facilitate timely information exchange, synchronize development timelines, promote knowledge sharing and collaboration, manage changes and dependencies, and implement continuous integration and testing practices of the product increments [65]. These factors play a critical role in ensuring that the integration process runs smoothly and efficiently. Therefore, it is imperative to prioritize these aspects and allocate adequate resources toward their implementation.

The solution proposed by this chapter is the integration of DT into an APM framework, such as Scrum. This integration is illustrated in **Figure 2**. The benefits of employing a DT in the complex levels of APM for the development of physical products are manifold. We will expound upon these advantages in the ensuing section.

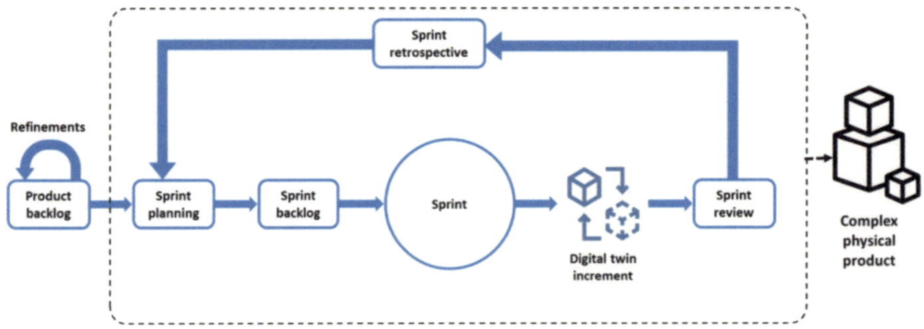

Figure 2.
Scrum framework using DT for complex physical product development.

By using this technology throughout the project, one can take a proactive approach toward addressing the challenges that might arise in complex projects despite the proper use of an agile approach.

4.1 Digital twin: a solution for managing complexity in agile project management of physical products

In the realm of complex physical product development projects, DT technology assumes an essential position, particularly when multiple teams are engaged in the development of diverse subsystems [66]. In complex projects, teams are often frequently geographically dispersed [67]. Consequently, the collaboration process, as well as the combination of the evolution of their subproducts and the results of testing from prototyping, can become even more challenging. However, it is worth noting that the integration process can be significantly streamlined through the use of DT technology. By leveraging the virtual representation of the physical product or subsystem, teams can effectively collaborate and exchange information regarding the project's evolution. This approach enables seamless teamwork and promotes efficient communication among team members [11].

The use of DT technology in APM enables each team to develop its specific subproduct independently, with a concentrated emphasis on its designated areas of expertise. The premise at hand is that the DT functions as a shared point of reference, facilitating cooperation and integration among the teams in question [68]. DT technology offers a collaborative platform that enables teams to visualize, analyze, and interact effectively with their respective subproducts. The use of DT technology fosters a sense of alignment and facilitates the integration process, which is a significant source of risk in complex projects. This way of working is shown schematically in **Figure 3**. The implementation of DT technology will enable teams to integrate their subproducts seamlessly in a virtual environment, facilitating the analysis of multiple interfaces and the refinement of their product backlogs.

By means of the integration that DT technology facilitates, agile teams are able to gain a comprehensive view of the entire scenario, even when they are geographically dispersed across different regions of the globe. Every team has the opportunity to make a valuable contribution to the DT by providing its respective subproduct, thereby ensuring that all components are seamlessly integrated. Virtual integration eliminates the necessity for physical conveyance of prototypes, diminishes temporal delays, and facilitates instantaneous collaboration [11]. The integration of

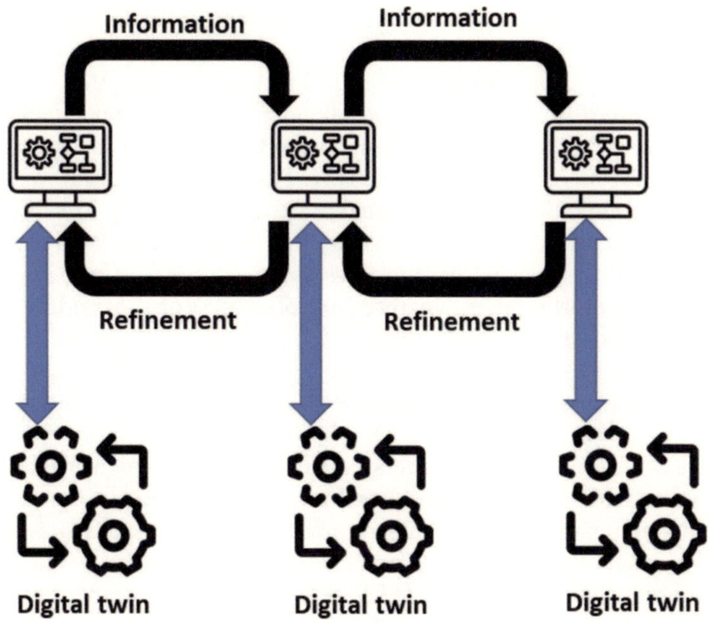

Figure 3.
Several agile teams exchanging information and refining their respective product via DT.

collaboration tools and communication channels within the DT platform serves to strengthen teamwork among geographically dispersed teams. Team members are able to engage in effective communication, exchange valuable feedback, and arrive at mutually beneficial decisions through the employment of the virtual platform.

DT technology facilitates the remote monitoring and control of physical entities, even in geographically dispersed locations, particularly in complex projects. Project teams have the ability to conduct remote assessments of performance, detect any anomalies that may arise, and take the appropriate actions as needed. The implementation of remote monitoring and control systems has proven to be a highly effective means of enhancing operational efficiency, minimizing the necessity for physical presence, and facilitating prompt responses to any arising issues [69]. Furthermore, DT technology offers advantages that transcend the development phase because it continues to provide value throughout the entire product lifecycle.

In essence, the use of DT technology within the context of complex APM of physical products yields significant benefits to project coordination. The project's success is achieved through the facilitation of iterative design and testing processes as well as the rapid synchronization of teams involved in module prototyping and simulations. Through the development of virtual prototypes of tangible entities and the subsequent exploration of diverse design alternatives in a virtual environment, the necessity for physical prototypes is significantly diminished. The development team has the opportunity to use integrated real-time data from a multitude of sources, including but not limited to sensors, IoT devices, and monitoring systems [70]. The capability to share data guarantees that additional stakeholders, including clients, are granted access to the latest and most accurate information. In complex projects, the exchange of real-time data diminishes the possibility of operating with obsolete or contradictory information, thereby augmenting project efficacy and agility. Henceforth, DT not only serves as a resolution to the numerous specific obstacles encountered during the development of physical products, but it also has the potential to further improve APM across various knowledge domains of project management.

To summarize, the implementation of a DT in an agile project that involves numerous geographically dispersed teams collaborating on a complex new product yields several benefits. The provision of a shared reference point for all stakeholders facilitates collaboration and knowledge sharing. The DT's capacity for real-time functionality facilitates expeditious decision-making and adaptability. The use of virtual testing and experimentation has proven to be a highly effective means of reducing costs and expediting the development process. Furthermore, the implementation of DT technology elevates the general quality and robustness of the product, resulting in heightened levels of customer satisfaction.

4.2 Enhancing agile project performance with digital twin technology

As we have demonstrated, APM offers a comprehensive framework for effectively tackling the intricacies of complex projects. Moreover, it is noteworthy that the DT constitutes a virtual platform that enables teams to combine their efforts, gain a comprehensive visualization of the entire system, and conduct integration testing in a virtual environment. The integration of DT technology with complex physical product projects can yield numerous benefits for project management. For instance, by using virtual representations, stakeholders can be more effectively engaged, project scope can be managed with greater efficiency, product quality can be ensured, costs can be optimized, delays can be minimized, and technical risks can be mitigated.

The concept of DT presents a remarkable opportunity for collaborative efforts in project management. This innovative solution proves particularly advantageous in the development of complex physical products, especially when dealing with teams that are geographically dispersed. By facilitating the integration of subproducts, DT technology enables seamless collaboration among team members, thereby enhancing the overall efficiency of the project. This particular feature enables the product owner, in particular, to examine and deliberate upon the entity from various angles, thereby enhancing lucidity and minimizing ambiguity in conversations with customers.

DT technology offers a virtual replica of complex products, allowing stakeholders to seamlessly engage with and visualize the product in a realistic and intuitive manner [71]. This visualization serves as a valuable tool for stakeholders because it enables them to gain a comprehensive understanding of the product's behavior, capabilities, and potential challenges. This, in turn, facilitates informed decision-making, which is crucial for the success of the project. Stakeholders are encouraged to engage actively in the development process by providing constructive feedback, offering insightful suggestions for improvement, and ensuring that their unique requirements and expectations are effectively addressed. Such active participation is highly valued by most stakeholders and can significantly enhance the overall quality of the development process. The use of DT technology promotes cooperation and cultivates a shared sense of responsibility among stakeholders, ultimately leading to enhanced project results and heightened stakeholder satisfaction.

The optimization of product design is a crucial objective in the development of complex products. In this regard, DT presents a potent platform that can be leveraged to attain this goal. Using the virtual environment, teams have the opportunity to examine various design iterations thoroughly, evaluate their efficacy, and ultimately make informed design decisions based on data-driven insights [72]. The capacity to iterate expeditiously within the digital realm enables the attainment of more efficient and effective design optimization, culminating in products that surpass or meet customers' expectations.

The implementation of DT technology offers a valuable opportunity for teams to examine various design alternatives thoroughly. This process allows for the identification of the optimal design configuration that effectively balances performance, cost, and other critical factors. The use of DT technology in the optimization of product design serves to augment the overall competitiveness and marketability of the end product [73]. Furthermore, the use of DT technology empowers teams to validate the functionality of a product, optimize its performance, and ensure that it meets the desired quality standards.

In the same token, in the realm of product design, quality assurance is an indispensable element of complex product development projects. In this regard, DT technology plays an essential part in guaranteeing the quality of the product. DT technology facilitates a virtual environment that allows for comprehensive testing, simulations, and performance evaluations [11]. This innovative approach empowers teams to monitor the behavior and performance of the product closely, thereby enhancing the overall quality of the result of the project. The use of DT in APM facilitates the prompt identification and rectification of any design deficiencies, inadequacies, or operational hindrances.

It is also important to note that in the realm of project management, it is frequently the case that the overall success of a given endeavor is contingent upon the expeditiousness with which it is brought to fruition [74]. The ramifications of project timeline delays can be quite substantial, particularly in the realm of complex product development. However, DT technology can be instrumental in mitigating such delays by enabling teams to swiftly execute iterations, simulations, and scenario planning [11, 75]. By simulating various scenarios and evaluating their effects on project timelines, teams can effectively identify and address potential delays in the development process.

In addition, it is worth noting that APM's collaborative and transparent approach, when coupled with the remarkable abilities of DT technology, promotes seamless communication and alignment among all team members and stakeholders, which helps mitigate any potential delays that may arise as a result of miscommunication or misalignment. In the case of complex product development, effective cost control is an essential component that transcends only timely completion. In this regard, DT technology presents a host of significant benefits. The use of DT's virtual environment empowers teams to effectively optimize resource allocation and efficiently identify opportunities for cost-saving measures. Through the process of simulating diverse scenarios, teams are able to evaluate the effects of multiple factors on costs, thereby enabling them to make informed decisions that optimize resource use. The use of DT technology also enables effective collaboration and communication among teams that are geographically separated, thereby diminishing the necessity for physical presence and the accompanying travel expenses [11, 73]. Another important benefit of the use of DT in APM is that the capacity to identify and rectify design deficiencies or performance inadequacies in the initial stages of the development process serves to mitigate the expenses associated with extensive rework or redesign.

It is imperative to remember that complex product development naturally entails technical risks [76, 77]. However, the use of DT technology also effectively alleviates these risks. The DT platform offers a virtual environment that enables teams to create models and simulate complex technical systems. This allows for the identification of potential issues, evaluation of various design alternatives, and informed decision-making to mitigate risks. The implementation of a proactive risk management approach can effectively decrease the probability of encountering technical setbacks, rework, or failures, thereby augmenting the project's overall triumph. When managing complex projects, DT with APM entails more than just basic risk management. In fact, it incorporates learning and knowledge management into the mix. As teams engage with DTs, they acquire valuable insights and knowledge regarding the product and its behavior. The insights are subsequently shared among the organization's members via DT. This process of knowledge transfer not only facilitates the dissemination of valuable information but also nurtures a culture of perpetual learning within the organization [78, 79].

The knowledge and insights gained from past projects can be effectively used in upcoming projects, resulting in enhanced levels of efficiency and efficacy. The use of DT as a knowledge repository enables teams to conveniently access and capitalize on prior experiences, thereby mitigating the possibility of reiterating errors and fostering the adoption of optimal methodologies. The implementation of DT technology enables organizations to bolster their capabilities and expertise through the facilitation of learning and knowledge management, which leads to improved project outcomes and heightened competitiveness in the market.

In summary, the mutually beneficial connection between APM methodologies applied to complex product development and the use of DT technology presents significant advantages for businesses. APM is a highly adaptable and iterative approach that enables swift responsiveness to evolving customer requirements and market dynamics. DT technology offers a virtual depiction of the tangible product, facilitating continuous monitoring, experimentation, and evaluation during the entire duration of the product's existence. The amalgamation of these two fundamental concepts engenders a robust framework that enables the optimization of product development while concurrently augmenting customer satisfaction. Through the use of DT technology in APM, companies can elevate project results, produce superior products, streamline expenses, diminish delays, alleviate technical risks, foster knowledge acquisition, and stimulate innovation in complex product development initiatives.

5. Conclusion

The implementation of APM methodologies for the development of new physical products necessitates tailored adaptations to effectively tackle the unique challenges inherent in this particular domain. To attain success, it is imperative to tailor the product backlog, cultivate cross-functional collaboration, and underscore the significance of prototyping. By integrating prototypes at various phases of the development cycle, teams can authenticate designs, amass feedback, and progressively enhance the tangible product to fulfill user demands and market standards. The use of prototyping as a tool is highly valuable in the realm of product development. It serves to mitigate risks, reduce uncertainty, and effectively deliver innovative physical products that cater to the needs of the customer. The emergence of the DT concept has proven to be a revolutionary approach in bridging the gap between the physical and digital worlds. A DT refers to a computer-generated model that replicates or represents a physical object, process, or system in a virtual environment. The concept in question encompasses not only the tangible characteristics but also the behavior and conduct exhibited by what it denotes. Through the integration of real-time data, simulation, and analytics, DT technology offers a holistic comprehension and visualization of the physical entity in question. The integration of DT technology within the framework of APM presents a multitude of noteworthy benefits and prospects. The agile methodology for project management flourishes through the sustained cooperation and intimate engagement of cross-functional teams via DT.

The implementation of DT technology serves to enhance collaboration among diverse teams with varying areas of expertise in a given project. Through the integration of physical prototypes with their digital counterparts, designers are able to monitor and analyze performance metrics effectively, detect anomalies, and optimize the design based on real-world data. The DT functions as a universal point of reference for all parties involved, promoting cooperation and guaranteeing that all individuals possess a mutual comprehension of the product's prerequisites and specifications. The ongoing exchange of feedback between the physical prototype and its DT serves as an iterative process that propels perpetual enhancement and ingenuity. By using a common digital representation, professionals such as designers, engineers, and stakeholders can effectively collaborate from remote locations. This collaboration is made possible through the use of real-time data and simulations, which allow for seamless communication and exchange of ideas. The implementation of this approach facilitates efficient and seamless communication, swift feedback loops, and enhanced decision-making capabilities, even in situations where team members are geographically dispersed. The capability to tackle concerns promptly in the digital domain facilitates the streamlining of the design and development process, thereby diminishing the time and expenses linked with interface integration in complex projects, in contrast to instances where solely conventional physical prototyping is employed.

Furthermore, it is worth noting that the implementation of a DT has a significant impact on the product's general quality and dependability. By using the DT at every stage of the development process, teams can effectively and proactively anticipate and rectify any potential design flaws or performance bottlenecks. The capacity to conduct virtual testing and optimization of a product empowers teams to fine-tune and augment its functionality, thereby guaranteeing that it satisfies the intended specifications and requisites. Ultimately, this leads to the production of a superior product that is more closely attuned to customers' expectations. The incorporation of a DT into APM enables agile teams to gain insight into a variety of design alternatives, conduct material experiments, evaluate performance characteristics, and verify functionalities remotely through the DT.

To summarize, the harmonious collaboration between APM and the use of DT technology presents a multitude of benefits for enterprises engaged in the development of complex products. The amalgamation of agile principles with DT capabilities facilitates a collaborative environment, real-time monitoring, virtual testing, and customer engagement. The implementation of this approach amplifies the flexibility, swiftness, and excellence of the product development process, culminating in the creation of products that are more adept at satisfying customers' requirements and providing substantial worth. To conclude, the incorporation of DT technology into APM presents a revolutionary approach to tackling issues related to communication and information. Through the adoption of this collaborative approach, enterprises can maintain their competitiveness in a swiftly evolving marketplace and simultaneously propel innovation and elevate customer satisfaction to unprecedented levels.

Author details

Alencar Bravo* and Darli Vieira
Department of Management, Université du Québec à Trois-Rivières,
Trois-Rivières, QC, Canada

*Address all correspondence to: alencar.soares.bravo@uqtr.ca

References

[1] Brady T, Davies A. From hero to hubris – Reconsidering the Project Management of Heathrow's terminal 5. International Journal of Project Management. 2010;**28**:151-157

[2] Sytnyk H, Hanna S, Blazhenko S. Controlling of Operational Activity of Trade Enterprises. Ukraine: Herald of Kyiv National University of Trade and Economics; 2022

[3] Zhou Z, Wang X. Discussions on Project Management and its success: A comprehensive theoretical review. Scientific and Social Research. 2022;**4**:30-35

[4] Anwar S, Tistogondo J. Making use of the fast-track method for project work control analysis. Adri International Journal of Sciences Engineering and Technology. 2022;7:18-27

[5] Lie FN, Edy E. Analysis and Design of web-Based Project Management Information System Using Critical Path Method (CPM) Method on PT. Pakarti Tirtoagung. Indonesia: Tech-E; 2019

[6] Antypenko N, Lozhachevska O, Velychko O, Melnyk O. Project Management of innovation-oriented Organisations in the context of financial risks insurance, decentralisation policies and the sustainable development of a region. State and regions. Series Economics and Business. 2022;**1**:106-111

[7] Žužek T, Gosar Ž, Kušar J, Berlec T. Adopting agile Project Management practices in non-software SMEs: A case study of a Slovenian medium-sized manufacturing company. Sustainability. 2020;**12**:3245

[8] Azanha A, Argoud ARTT, Camargo JLV, Antoniolli PD. Agile Project Management with scrum. International Journal of Managing Projects in Business. 2017;**10**:121-142

[9] Erickson JS, Lyytinen K, Siau K. Agile modeling, agile software development, and extreme programming. Journal of Database Management. 2005;**16**(4):88-99

[10] Shankarmani R, Pawar R, Mantha SS, Babu V. Agile methodology adoption: Benefits and constraints. International Journal of Computer Applications. 2012;**58**:31-37

[11] Madni AM, Madni CC, Lucero S. Leveraging digital twin Technology in Model-Based Systems Engineering. Systems. 2019;7:7

[12] Gupta M, Khan N. Digital twin understanding, current progressions, and future perspectives. In: Mishra BK editor. Handbook of Research on Applications of AI, Digital Twin, and Internet of Things for Sustainable Development. IGI Global; 2023. pp. 332-343. DOI: 10.4018/978-1-6684-6821-0.ch019

[13] Akanmu A, Anumba CJ, Ogunseiju O. Towards next generation cyber-physical systems and digital twins for construction. Journal of Information Technology in Construction. 2021;**26**:505-525

[14] Xu X, Liu C, Wang KKW, Huang H, Xu X. Digital twin-driven smart manufacturing: Connotation, reference model, applications and research issues. Robotics and Computer-Integrated Manufacturing. 2020;**61**:101837

[15] Srivastava P, Jain S. A leadership framework for distributed self-organized scrum teams. Team Performance Management. 2017;**23**:293-314

[16] Rubin KS. Essential Scrum: A Practical Guide to the most Popular Agile

Process. United States: Addison-Wesley; 2012

[17] Vargas DAD, Baron C, Esteban P, Estrada CYAG. Is there any agility in systems engineering? Insight. 2017;**20**:11-14

[18] Hoda R, Murugesan LK. Multi-level agile Project Management challenges: A self-organizing team perspective. Journal of Systems and Software. 2016;**117**:245-257

[19] Mahmoud-Jouini SB, Midler C, Silberzahn P. Contributions of design thinking to Project Management in an innovation context. Project Management Journal. 2016;**47**:144-156

[20] Shenhar AJ, Dvir D, Stefanovic J. The myth and reality of Project Management. In: 2008 Portland International Conference on Management of Engineering & Technology. Cape Town, South Africa: IEEE; 2008

[21] Hoda R, Salleh N, Grundy J. The rise and evolution of agile software development. IEEE Software. 2018;**35**:58-63

[22] Beck K, Beedle M, Van Bennekum A, Cockburn A, Cunningham W, Fowler M, et al. The agile manifesto. The Agile Alliance. 2001. Available from: www.agilemanifesto.org

[23] Su SH, Scharff C. Know yourself and beyond: A Students' global software development project experience with agile methodology. In: 2010 5th International Conference on Computer Science & Education. Hefei, China: IEEE; 2010

[24] Nuottila J, Aaltonen K, Kujala J. Agile Project Management in a public context: Case study on forms of Organising. International Journal of Project Organisation and Management. 2017;**9**:230-248

[25] Özkan DÇ, Mishra A. Agile Project Management tools: A brief Comprative view. Cybernetics and Information Technologies. 2019;**19**:17-25

[26] Rodríguez P, Haghighatkhah A, Lwakatare LE, Teppola S, Suomalainen T, Eskeli J, et al. Continuous deployment of software intensive products and services: A systematic mapping study. Journal of Systems and Software. 2017;**123**:263-291

[27] Noteboom C, Ofori M, Sutrave K, El-Gayar OF. Agile Project Management: A systematic literature review of adoption drivers and critical success factors. In: Proceedings of the 54th Hawaii International Conference on System Sciences. 2021. pp. 6775-6785. Available from: https://scholarspace.manoa.hawaii.edu/server/api/core/bitstreams/37cf9dea-7864-4a93-ab68-b52e44488e53/content

[28] Chang M. An agile approach to library IT innovations. Library Hi Tech. 2010;**28**:672-689

[29] Scrum inc. 2022 State of Agile Report - 5 Takeaways 2022. Available from: https://www.scruminc.com/2022-state-agile-report-takeaways/

[30] Kuhl JG. Incorporation of Agile Development Methodology into a Capstone Software Engineering Project. Iowa: The University of Iowa; 2014

[31] Rachman NT, Sushandoyo D. Analysis of scrum implementation in digital startup product development. Journal of Economic Bussines and Accounting (Costing). 2021;**5**(1):190-196

[32] Alsohybe NT, Sabrah N. Agile Project Management in non-software sectors during turbulent times. International Journal of Software Engineering & Applications. 2019;**10**:51-68

[33] Soueid MI, Martins AFC. Scrum and agility beyond it: Evidences in the

Brazilian mining industry. Revista De Gestão E Projetos. 2021;**12**:123-148

[34] Maruping LM, Venkatesh V, Agarwal R. A control theory perspective on agile methodology use and changing user requirements. Information Systems Research. 2009;**20**:377-399

[35] Kazar G, Almhamdawee A, Tokdemir OB. Potential benefits of agile Project Management in improving construction project performances: A case study of Iraq. Journal of Construction Engineering Management & Innovation. 2022;**5**:64-76

[36] Neyens I, Faems D. Exploring the impact of Alliance portfolio management design on Alliance portfolio performance. Managerial and Decision Economics. 2013;**34**:347-361

[37] Adkar P, Shinde SG. A modern review on scrum: Advance Project Management method. International Journal of Trend in Scientific Research and Development. 2018;**2**:87-95

[38] Sutherland J, Sutherland J. Scrum: The Art of Doing Twice the Work in half the Time: Currency. New-York, United States: Random House Business Books; 2014

[39] Sverrisdottir HS, Ingason HT, Jonasson HI. The role of the product owner in scrum-comparison between theory and practices. Procedia-Social and Behavioral Sciences. 2014;**119**:257-267

[40] Shafiq S, Hafeez Y, Ali S, Iqbal N, Jamal M. Towards scrum based agile framework for global software development teams. Mehran University Research Journal of Engineering and Technology. 2019;**38**:979-998

[41] Noll J, Razzak MA, Bass JM, Beecham S. A study of the scrum Master's role. In: Felderer M, Méndez Fernández D, Turhan B, Kalinowski M, Sarro F, Winkler D, editors. Product-Focused Software Process Improvement. PROFES 2017. Vol. 10611. Cham: Springer; 2017

[42] Schwaber K, Beedle M. Agile Software Development with Scrum. USA: Prentice Hall PTR; 2001

[43] Ashari I, Aryani AF, Ardhi A. Design and build inventory management information system using the scrum method. Jsii (Jurnal Sistem Informasi). 2022;**9**:27-35

[44] Hron M, Obwegeser N. Why and how is scrum being adapted in practice: A systematic review. Journal of Systems and Software. 2022;**183**:111110

[45] Ferguson S, Lai K, Chen JM, Faidi S, Leonardo K, Olechowski A. "Why Couldn't we do this more often?": Exploring the feasibility of virtual and distributed work in product design engineering. Research in Engineering Design. 2022;**33**:413-436

[46] Kantaros A, Soulis E, Ganetsos T, Petrescu FIT. Applying a combination of cutting-edge industry 4.0 processes towards fabricating a customized component. PRO. 2023;**11**:1385

[47] Zorriassatine F, Wykes C, Parkin RM, Gindy N. A survey of virtual prototyping techniques for mechanical product development. Proceedings of the Institution of Mechanical Engineers Part B Journal of Engineering Manufacture. 2003;**217**:513-530

[48] Blindheim J, Elverum CW, Welo T, Steinert M. Concept evaluation in new product development. Journal of Engineering Design and Technology. 2020;**18**:1139-1151

[49] Shalpegin T, Sommer SC, Wan Z. Collaborative prototyping of alternative

designs under a target costing scheme. Production and Operations Management. 2017;**27**:496-515

[50] Schork S, Kirchner E. Purpose-oriented modelling of the learning process when using prototypes. Proceedings of the Design Society International Conference on Engineering Design. 2019;**1**:1573-1582

[51] Williams L, Brown G, Meltzer AB, Nagappan N. Scrum + engineering practices: Experiences of three Microsoft teams. In: 2011 International Symposium on Empirical Software Engineering and Measurement. Banff, AB, Canada: IEEE; 2011

[52] Suomela R, Koskinen K, Heikkinen K. Rapid prototyping of location-based games with the multi-user publishing environment application platform. In: The IEEE International Workshop on Intelligent Environments. Colchester, UK: IET; 2005

[53] Sultana J. A Study on Applicability of the Scrum Framework for Large Software Projects [Thesis]. Canada: Ryerson University; 2021

[54] Mohamed O, Nour M, Elazizy I, Hagras M. A suggested module for risk framework of water control construction projects using qualitative and quantitative techniques. International Journal of Engineering and Advanced Technology. 2020;**9**:1448-1454

[55] Raydugin YG. Modern Risk Quantification in Complex Projects: Non-linear Monte Carlo and System Dynamics Methodologies. Oxford University Press; 2020

[56] Awotunde JB, Ayo FE, Ogundokun RO, Matiluko OE, Adeniyi EA. Investigating the roles of effective communication among stakeholders in collaborative software development projects. In: Gervasi O et al., editors. Computational Science and its Applications – ICCSA 2020. Vol. 12254. Cham: Springer; 2020

[57] Ju Q-q, Ding L, Skibniewski MJ. Optimization strategies to eliminate Interface conflicts in complex supply chains of construction projects. Journal of Civil Engineering and Management. 2017;**23**:712-726

[58] Galli BJ. The value of communication Management in Agile Project Environments. International Journal of Applied Logistics. 2022;**12**:1-21

[59] Malik S, Taqi M, Martins JM, Mata MN, Pereira J, Abreu A. Exploring the relationship between communication and success of construction projects: The mediating role of conflict. Sustainability. 2021;**13**(8):4513

[60] Belogrudova MZ, Antonova L. Correlation of internal communication with emotional intelligence in project teams in organization "X". Society Integration Education Proceedings of the International Scientific Conference. 2021;**6**:540-550

[61] Kakkar A, Duggal P. Investigating the factors affecting delays in infrastructure projects. International Journal of Engineering and Management Research. 2021;**11**:273-280

[62] Mundra A, Misra S, Dhawale CA. Practical scrum-scrum team: Way to produce successful and quality software. In: 2013 13th International Conference on Computational Science and its Applications. Vietnam, New-York, United States: IEEE; 2013

[63] Zhu F, Hu H, Xu F. Risk assessment model for international construction projects considering risk interdependence using the DEMATEL method. PLoS One. 2022;**17**(5):e0265972

[64] Bai L, Zhang K, Shi H, An M, Han X. Project portfolio resource risk assessment considering project interdependency by the fuzzy Bayesian network. Complexity. 2020;**2020**:1-21

[65] Eslami MH, Lakemond N. Knowledge integration with customers in collaborative product development projects. Journal of Business and Industrial Marketing. 2016;**31**:889-900

[66] Ahmad S, Mallick DN, Schroeder RG. New product development: Impact of project characteristics and development practices on performance. Journal of Product Innovation Management. 2012;**30**:331-348

[67] Hoegl M, Ernst H, Proserpio L. How teamwork matters more as team member dispersion increases. Journal of Product Innovation Management. 2007;**24**:156-165

[68] Aretoulaki E, Ponis ST, Plakas G. Complementarity, interoperability, and level of integration of humanitarian drones with emerging digital technologies: A state-of-the-art systematic literature review of mathematical models. Drones. 2023;**7**(5):301

[69] Tekinerdogan B. On the notion of digital twins: A modeling perspective. Systems. 2022;**11**(1):15

[70] Kurvinen E, Kutvonen A, Ukko J, Khadim Q, Hagh YS, Jaiswal S, et al. Physics-based digital twins merging with machines: Cases of Mobile log crane and rotating machine. IEEE Access. 2022;**10**:45962-45978

[71] Kerin M, Pham DT, Huang J, J H. A generic asset model for implementing product digital twins in smart remanufacturing. The International Journal of Advanced Manufacturing Technology. 2022;**124**:3021-3038

[72] Hämäläinen M. Urban development with dynamic digital twins in Helsinki City. Iet Smart Cities. 2021;**3**:201-210

[73] Li Q, Li Z, Tang X, Yang H, Song Y. Demonstration and validation of the digital twin Technology for a Regional Multi-Energy System. Proceedings of the SPIE. 2023;**12645**:126450U

[74] Varajão J, Marques RP, Trigo A. Project Management processes – Impact on the success of information systems projects. Informatica. 2022;**33**(2):421-436

[75] Athavale VA. Digital twin - a key technology driver in industry 4.0. Engineering Technology Open Access Journal. 2021;**4**(1):555628

[76] Maruping LM, Venkatesh V, Thong JYL, Zhang X. A risk mitigation framework for information technology projects: A cultural contingency perspective. Journal of Management Information Systems. 2019;**36**:120-157

[77] Osman K, Stamenkovic D, Lazarević MP. Integration of system design and production processes in robust mechatronic product architectures development - extended M-FBFP framework. Hemijska Industrija. 2013;**67**:759

[78] Merlo TR. Knowledge management technology: Human-Computer Interaction & Cultural Perspective on pattern of retrieval, organization, use, and sharing of information and knowledge. Knowledge and Performance Management. 2017;**1**(1):46-54

[79] Migdadi MM. A knowledge-centered culture as an antecedent of effective knowledge Management at Information Technology Centers in the Jordanian universities. Journal of Systems and Information Technology. 2009;**11**(2):89-116

Section 3

Applications of Digital Twins

Chapter 4

Perspective Chapter: Digital Twins for Health – Opportunities, Barriers and a Path Forward

Patrizio Armeni, Irem Polat, Leonardo Maria De Rossi, Lorenzo Diaferia, Giacomo Visioli, Severino Meregalli and Anna Gatti

Abstract

The concept of precision medicine involves tailoring medical interventions to each patient's specific needs, considering factors such as their genetic makeup, lifestyle, environment and response to therapies. The emergence of digital twin (DT) technology is anticipated to enable such customization. The healthcare field is, thus, increasingly exploring the use of digital twins (DTs), benefiting from successful proof of concept demonstrated in various industries. If their full potential is realized, DTs have the capability to revolutionize connected care and reshape the management of lifestyle, health, wellness and chronic diseases in the future. However, the realization of DTs' full potential in healthcare is currently impeded by technical, regulatory and ethical challenges. In this chapter, we map the current applications of DTs in healthcare, with a primary focus on precision medicine. We also explore their potential applications in clinical trial design and hospital operations. We identify the key enablers of DTs in healthcare and discuss the opportunities and barriers that foster or hinder their larger and faster diffusion. By providing a comprehensive view of the current landscape, opportunities and challenges, we aim to contribute to DTs' ongoing development and help policymakers facilitate the growth of DTs' application in healthcare.

Keywords: human digital twins, precision medicine, DT of the eye, clinical trials, hospital operations management

1. Introduction

Digital twins (DTs) are currently implemented in various sectors such as smart cities, manufacturing, construction, automotive and aerospace. Successful proof of concept in various industries and the growing accessibility of technological devices capable of gathering patient data facilitate an emerging utilization of DTs in the healthcare field [1]. DTs present significant potential, particularly in the realm of precision medicine [2]. In this context, DTs offer the capability to simulate personalized treatment approaches, providing a visual representation of the potential outcomes of therapies and the progression of diseases for each individual patient [3]. Furthermore, the healthcare sector has already witnessed successful implementations of DTs, particularly in areas like predictive maintenance and performance enhancement of medical devices, as well as the optimization of hospital management systems.

While DTs have demonstrated significant success in various fields, it is important to acknowledge that DTs developed for healthcare differ significantly from those designed for industrial applications. Healthcare DTs face unique challenges and requirements due to the complexity and sensitivity of medical data, and there are several important points to consider. For instance, human DTs indeed rely heavily on the integration of artificial intelligence (AI) applications, and these applications often utilize sensitive medical data. Given the sensitive nature of medical data, ensuring the proper handling and protection of this information becomes crucial [4].

Unlike other industrial applications, health DTs have unique characteristics that require a dedicated examination. This review encompasses an evaluation of the technologies used in health DTs, the specific areas of application within healthcare, and the barriers that exist in both the research field and the market for these technologies. By conducting this comprehensive analysis, a deeper understanding can be gained regarding the potential benefits, challenges and opportunities associated with health DTs in healthcare settings.

This chapter contributes to the DTs in healthcare literature as follows:

- We introduce a comprehensive background and recent advances of DTs in healthcare covering multiple areas of healthcare.

- We provide a comprehensive summary of enabling technologies and possible data sources in DTs in healthcare, assessing their benefits and drawbacks.

- We summarize DT applications in healthcare (precision medicine, clinical trials design and hospital operations) from the literature. As a case study, we present a comprehensive review focused on DTs of the eye and how they differ from creating DTs of other organs.

- Finally, we discuss limitations and open issues. We engage in a discussion on potential strategies to address these challenges and overcome the barriers they pose.

2. Background: concept of DTs in healthcare

A Digital Twin (DT) serves as a reflection of the real world, offering a way to simulate, predict and enhance physical manufacturing systems and procedures [5]. In the realm of healthcare, DTs can be described as virtual representations, also known as "digital twins," of patients, their anatomical structure and medical devices. These twins are created using a combination of multimodal patient data, population data and real-time updates on patient and environmental variables [6, 7]. Additionally, when it comes to healthcare facilities, a DT of a hospital environment can also be included within the scope. Overall, in the healthcare field, DTs offer various benefits, covering diverse areas such as enhancing diagnostics, treatment and care, assisting in medical pathway planning, as well as supporting hospital organization and management, and facilitating medical or asset resource allocation.

The concept of human DTs comprises three essential components: the physical object, the virtual object and the digital thread. The author [8] summarizes these three components as follows: (1) The physical object can be a patient, a medical device, a wearable device, an external factor (e.g., social behavior, weather, air quality, or even government policies influencing patient health), or a system consisting of more of these objects (e.g., a hospital). (2) The virtual object is the medical device model, wearable device model, digital person model, external factor model and digital system models. (3) The digital thread is healthcare data, including real-time data detected from medical and wearable devices or external factors, simulation data from digital models, historical health data and electronic health records (EHRs) from healthcare institutions, and service data from platforms that enable the communication between the physical and virtual objects and spaces.

The physical object in DT applications primarily comprises physical patients and devices such as medical instruments, auxiliary equipment and wearable sensors that are connected to actual individuals. Various medical detection and scanning instruments, as well as wearable devices, are utilized to collect dynamic and static multi-source data related to physical humans. This data is then transmitted in real-time to a virtual space and is processed to finally return to the physical objects as real-time instructions and commands (**Figure 1**). This interplay between the physical twin (PT) and virtual twin (VT) enables seamless interaction and data exchange for improved healthcare outcomes [8]. Thus, to enable the continuous development of DTs, the co-evolution of both PTs and VTs is essential. A reliable data link facilitates the mapping between the PT and VT, enabling their synchronized development in both the physical and virtual environments. This enables real-time data analysis and continuous monitoring of the PT's condition, allowing for instance the early detection or crisis warning of potential health issues [9].

DT models can be effectively employed to enhance the outcomes of diverse clinical procedures. By leveraging real-time data and processing capabilities, DT algorithms can provide accurate insights and predictions, thereby improving decision-making in

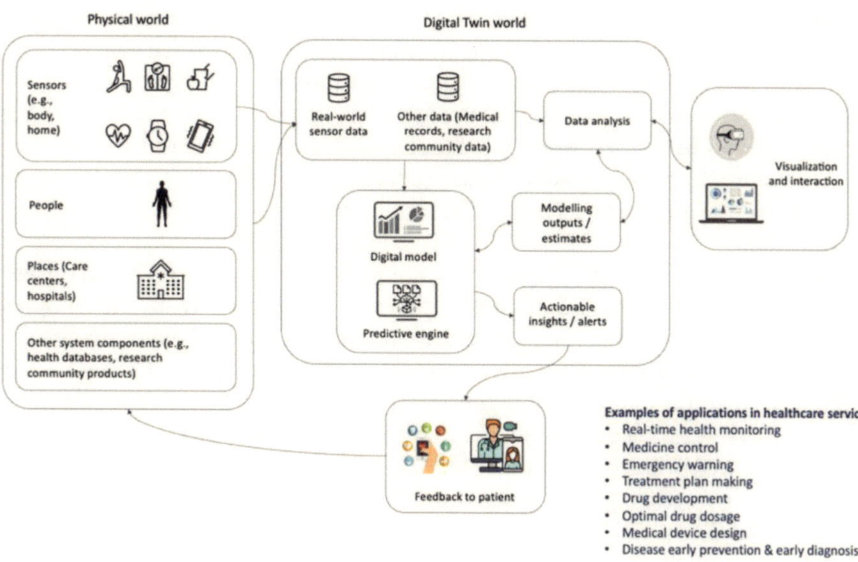

Figure 1.
High-level component view of a DT for precision medicine.

medical settings. Through the utilization of a vast dataset and AI-powered models, an ideal replica of the human body or specific body parts can be generated. This replication mimics human physiology and can provide potential answers to a wide range of clinical questions [10, 11]. Additionally, DTs can provide a secure environment for young practitioners, doctors and surgeons to practice, undergo training procedures and conduct tests on a virtual representation of the human body. This enables them to enhance their skills and expertise in a safe environment.

3. Health DT development: enabling technologies and data sources

The DT model allows for the continuous collection and accumulation of data throughout its entire lifecycle [12]. The accessibility of biomedical data has significantly increased due to various sources such as large biobanks, electronic health records, medical imaging, wearable devices, biosensors, as well as cost-effective genome and microbiome sequencing. This accessibility has played a crucial role in the development of health DT solutions [13]. Moreover, data gathered through personal digital devices, and patient-generated health data, such as patient-reported function or symptoms, physical markers, and demographic, and lifestyle data over time of an individual contribute to the development of more comprehensive DTs [9, 14, 15]. The integration of diverse data streams within health DTs allows for a broad understanding of an individual's health. This holistic view enables personalized healthcare interventions tailored to the specific needs of each individual.

In a health DT process, collected data from patients and the surrounding environment is transmitted and stored in real-time within the Internet of Things (IoT) cloud. Through the utilization of big data analytics and AI, valuable insights are extracted from the vast volume of data. This knowledge can be reused and enhanced over time. These insights then enable the creation of a VT, representing the PT's condition. Through this process, information regarding the PT's attributes, health status and other relevant data is fed back to the virtual models, enabling a two-way transmission of data. Thus, effective communication techniques that enable bi-directional data transmissions between the PT and VT are crucial for the success of this model. By constructing product models in the virtual space and facilitating the feedback of digital models to the physical space, a closed-loop process is achieved in the DT mode [9]. Finally, to comprehend and monitor the PT's status comprehensively, visualization tools are necessary. These tools allow for a visual representation of the PT's data, aiding in understanding and monitoring their condition.

The progress made in Digital Health technologies has made it possible to observe a rich amount of digital data and detailed aspects of people's behavior, the complex factors that influence behavior at any given moment, and how behavior changes over time within each person. This advancement relies on a range of devices such as smartphones, wearables, implantable sensors and ingestible sensors (like smart pills) to gather and analyze biological (e.g., blood glucose), physiological (e.g., heart rate and blood pressure) or behavioral data [16, 17]. Wearable sensors particularly enable healthcare professionals to gather real-time data outside of traditional clinical settings. The availability of affordable and noninvasive devices, such as smartwatches or bands, has rapidly increased. These wearable sensors are capable of accurately measuring various physiological metrics. By integrating data from these wearable devices with EHRs, it becomes possible to extract relevant information about a patient's underlying disease risk. This integration enables the creation of a personalized remote monitoring experience for patients and caregivers [13].

Moreover, DT models also rely on biomarkers that cannot be directly measured or require invasive procedures. These biomarkers play a crucial role in various applications, such as precision cardiology. In precision cardiology, the integration of cardiovascular imaging with computational fluid dynamics enables the noninvasive assessment of flow patterns and the computation of diagnostic metrics. This approach is particularly useful for conditions like coronary artery disease, aortic aneurysms, valve prostheses and stent design, providing valuable insights without the need for invasive procedures [18].

Despite these advancements in the healthcare field, there are several challenges (technical limitations, ethical considerations and financial constraints) that hinder data acquisition for building DTs and unfolding their full potential. Unlike industries like automotive, where sensors are readily available and integrated into the assets, humans do not naturally possess embedded sensors [3, 19]. Still, humans mostly rely on periodic medical examinations to gather data about their health; thus, the intermittent data collection method poses limitations in maintaining real-time and continuous updates for human DTs. The seamless connection between humans and their DTs cannot be guaranteed yet [4].

The implementation of DTs encompasses a wide range of technologies. These include the IoT, 5G networks, cloud and edge computing, extended reality (XR), simulation tools, visualization tools, and AI and machine learning (ML) models. Additionally, the integration of technologies such as federated learning (FL) and blockchain is rising to address security, transparency and privacy-related issues. These technologies coupled with the availability of diverse and accurate create exciting opportunities for the use of DTs in healthcare [3, 12, 19, 20].

3.1 Internet of Things

Enhanced IoT sensors and devices refer to internet-connected sensors and devices that have the capability to be integrated into ordinary objects or attached to the human body, such as wearable devices. These advanced sensors and devices enable data collection, communication and interaction with the surrounding environment, thereby enhancing connectivity and enabling a wide range of applications and services [21]. The integration of IoT into medical systems holds immense potential for driving the future of DTs in healthcare. Rather than relying solely on visits to hospitals, real-time health monitoring of patients enabled by IoT devices can empower individuals, allow early detection of health issues, and facilitate effective management of chronic conditions. This shift toward continuous monitoring and personalized healthcare plans has the potential to significantly improve patient outcomes and enhance overall healthcare delivery [22].

The decreasing cost and increasing accessibility of IoT devices have facilitated the rise of connectivity. However, challenges remain in achieving real-time monitoring, particularly due to factors like power outages, software errors and ongoing deployment issues. These challenges pose significant obstacles to the overall objective of establishing seamless connectivity. For instance, one of the challenges is the reliance on a single sensor to provide data for AI algorithms. If there is a dysfunction or failure in the sensor, it can hinder the success of a specific process. Achieving complete connectivity and redundancy in data collection is crucial to mitigate the impact of such failures.

3.2 5G network

The high transmission rate of the 5G network enables the collection of sensor data at a rate that satisfies the demands of big data analysis and advanced forms of AI in DT systems [23]. Thus, the adoption of 5G networks in DT applications has the potential to enhance ongoing operations through continuous monitoring of physical systems in real time. Despite this clear potential, DTs adoption within 5G networks remains relatively new [24].

3.3 Artificial intelligence

AI is revolutionizing healthcare by employing a widely applied combination of highly complex algorithms that emulate human cognitive functions across various applications and sectors. This transformative technology, including techniques like deep learning (DL) and ML, can be extensively applied to diverse healthcare data types [11, 22]. Multimodal AI models have the potential to integrate data from multiple sources, such as biosensors, genetic information, epigenetic markers, proteomic data, microbiome profiles, metabolomic measurements, medical imaging, textual data, clinical records, social determinants of health and environmental data [13].

DT applications leverage AI technologies and techniques, and software analytics to create and maintain a dynamic, real-time digital representation of a physical object. Currently, AI-powered DTs of human biological systems or organs play a significant role in diagnosing existing medical conditions and forecasting potential future health issues. This is accomplished by analyzing aggregated data and medical histories associated with individuals [25]. Furthermore, AI is crucial in designing DTs of organs by leveraging physiological data to generate 3D images. A notable example is the development of a DT model by Siemens Healthineers, which utilized a vast database comprising over 250 million annotated images, reports and operational data. By harnessing AI capabilities, this DT model enables the creation of digital heart designs based on patient-specific data, considering factors such as size, ejection fraction, muscle contraction and other relevant conditions unique to each patient. This AI-driven approach facilitates personalized and precise modeling of organs, contributing to advancements in healthcare diagnostics and treatment planning [26].

One of the major challenges DTs faces is particularly regarding privacy concerns. The collection and storage of large volumes of data in a centralized repository raise significant privacy concerns and increase the risk of data breaches. Solely removing patient identifiers from the data is not sufficient to address these concerns since there is a possibility of reconstructing the original data even without the identifiers. FL offers a potential solution to this issue by enabling the utilization of the combined power of individual data modalities without the need to centralize the data [27].

3.4 Federated learning

FL aims to address data governance and privacy concerns by enabling collaborative training of algorithms without sharing the actual data. FL allows institutions to gain insights collectively, such as through a consensus model, while keeping patient data within the institution. The ML process takes place locally at each participating institution, with only model characteristics (such as parameters and gradients) being transferred. Recent research has demonstrated that FL-trained models can achieve performance levels similar to those trained on centralized datasets and outperform models that only have access to isolated data from individual institutions [27, 28].

FL has the capacity to facilitate large-scale precision medicine, resulting in models that provide unbiased decisions, accurately represent an individual's physiology, and account for rare diseases, all while addressing governance and privacy concerns. However, the implementation of FL still requires careful technical considerations (including data heterogeneity, traceability and explainability issues) to ensure optimal algorithm performance without compromising safety or patient privacy [28].

3.5 Immersive technologies (XR)

The emergence of immersive technologies in both industrial and consumer electronics has introduced innovative possibilities for DTs. These technologies offer new paradigms that can enhance the visualization and interaction capabilities of DTs, and enable highly realistic simulations. The combination of DTs and immersive technologies has primarily been utilized in the manufacturing domain [29], and there is emerging evidence of its potential use cases in healthcare as well. Some use cases of VR in conjunction with DTs offers a safe and immersive platform for training, enhancing skills, and refining medical techniques, remote operations, and remote collaborations, for instance, in surgeries [30]. This integration allows clinicians to practice complex procedures in a virtual environment. For instance, [31] developed a novel DT prototype that facilitates remote surgeries by integrating a robotic arm and a VR system connected over a 4G mobile network. By testing the prototype, the authors were able to analyze communication and cybersecurity requirements within their DT system. Moreover, medical education benefits significantly from this advancement, as healthcare professionals can gain hands-on experience by practicing treatments and procedures on virtual patients before performing them on real individuals.

3.6 Cloud and edge computing

DTs rely on a substantial volume of data, requiring high computing power to enable clinicians to extract real-time patient information. However, the storage and computing capabilities required for DTs often surpass what is currently available in healthcare centers [32]. Thus, many healthcare centers outsource their healthcare data and monitoring services to different locations, such as the edge or the cloud.

The deployment decision of twins in healthcare is primarily influenced by two key factors: available computing power and latency. In a typical Cloud Computing (CC) setup, data storage and computation are carried out within a centralized system. Cloud deployment provides greater computing power but higher latency due to its remote nature [10, 33]. As the number of IoT devices, mobile services, and the size of data continue to grow rapidly, it becomes imperative to alleviate the computational burden on the operating station or cloud. As an alternative, Edge Computing (EC) enables the network to conduct computation or process data at the extreme edges of the network, closer to the data source, rather than relying on centralized or distributed nodes in the core of the network [10]. EC offers limited computing power compared to CC but also benefits from low latency as it is in close proximity to devices. In the healthcare domain, EC comes into play when time is critical when dealing with emergencies such as ischemic heart disease (IHD) or stroke, as it improves efficiency by reducing data circulation and providing faster data processing [34]. In 2019, [34] developed Cardio Twin, a platform designed as a DT of the human heart. The purpose of this DT platform is to detect, prevent and mitigate the risk of heart disease. Cardio Twin runs on the edge devices like smartphones and connects with external sensors through Bluetooth communication to gather biosignals and collects data from other sources like medical records. In turn, this data is processed to detect and help in case the real twin is suffering an IHD or a stroke.

The integration of AI and ML algorithms with EC will significantly contribute to the progress of various applications, including healthcare and industries. A novel concept called Edge Machine Learning enables smart devices to perform local processing utilizing ML and DL algorithms. While edge devices can still transmit data to the cloud, processing data locally offers several advantages. It allows for data screening before sending it to the cloud and facilitates real-time data processing and response [10].

Finally, an integrated cloud-edge computing framework will support the advancement of healthcare DTs by ensuring the availability of low-latency and high-capacity storage solutions. A cloud-edge computing arrangement enables time-sensitive tasks to be accomplished at the network's edge. Tasks that require heavy computation or storage and cannot be executed at the edge are transferred to the cloud [9]. Ultimately, it is crucial for an effective DT framework to establish mechanisms that guarantee security and privacy, ensure highly reliable communications and reduce latency [35]. The decision to deploy DTs involves a careful balance, considering the aforementioned characteristics [33].

3.7 Blockchain

Recent advancements in DTs could pose specific challenges (e.g., privacy and security) in data sharing, storage and access in the healthcare sector [36]. Blockchain technology could address them by storing a patient's medical history records in a secure, transparent, trustworthy and timely manner [37]. This ensures that healthcare providers and professionals have access to reliable and up-to-date information whenever needed. Having access to such comprehensive and timely data enables healthcare providers to make informed decisions and consider appropriate courses of action in the event of any future complications or medical concerns [38]. Despite the benefits of blockchain-enabled systems, one of the challenges they face is high latency caused by complex consensus mechanisms. This latency can hinder the system's efficiency in meeting DTs' low latency requirements. To overcome this challenge, novel optimization schemes are necessary [9].

4. Current applications of DTs in healthcare

In this section, we explore various examples of DT applications that have made significant advancements in facilitating the use of DTs in multiple healthcare domains. The primary objective of this section is to demonstrate the deployment of DTs in precision medicine and their role in supporting medical decision-making. Additionally, we introduce a detailed review of developing a DT of the eye. To provide a comprehensive overview of DTs in healthcare, we also present our findings on the utilization of DTs in clinical trial design and optimizing hospital operations.

4.1 Precision medicine and medical decision-making support

Precision medicine can be defined as an approach to target the right treatments to the right patients at the right time [39]. The broad goal of precision medicine is to deliver customized therapies to individual patients, aiming to optimize both the effectiveness of treatments and the overall efficiency of our healthcare system (e.g., to prevent disease, improve survival and extend health span) [18, 40]. However, the original concept of precision medicine has faced criticism for its heavy emphasis on genomics and its limited focus on addressing clinical management challenges [40]. Additionally, a significant barrier to achieving precision medicine is the lack of consistent treatment response among patients with the same disease. This discrepancy primarily results from the substantial complexity of the underlying condition, which can involve complex interactions among thousands of genes that vary across individuals with the same diagnosis. As a result, the concept is evolving and expanding to incorporate a broader range of data, including lifestyle factors, environmental influences and biological information, moving away from a solely gene-centric perspective [41].

Precision medicine necessitates not only improved and more comprehensive data but also advancements in computer capabilities to analyze, integrate and leverage this data, ultimately constructing a DT of an individual patient. In this context, DTs have the potential to facilitate the prediction of illnesses by analyzing the personal history of an individual's real twin and considering its current state, including factors such as location, time and activity. By leveraging the data collected from the real twin, DTs can simulate and predict the potential impact of different treatments on these patients, a shift from the "one-size-fits-all" treatments to tailor-made treatments [26]. This capability enables DTs to provide valuable insights into personalized treatment approaches, enabling them to make informed decisions and optimize patient care [18, 26]. An important challenge lies in integrating this data with healthcare organizations while ensuring the security and confidentiality of sensitive information [18].

In the medical and clinical fields, there is a growing interest and increasing availability of prototypes in the development of DTs in the precision medicine scope. The author in [42] proposed a framework for DT of patients, where a DT representing a patient exhibiting symptoms of a specific disease is created in unlimited copies, replicating the network models of all relevant molecular, phenotypic and environmental factors associated with the disease's mechanisms. Subsequent simulations are conducted using various drugs to determine the optimal treatment strategy.

DT technology is also encouraging to mimic human organs. The human heart [43], brain [44] and liver [45] are some examples of research areas within the DT scope. The Living Heart (Dassault Systèmes) project presented in [43] is a pioneering initiative in the field of organ DTs. This research project introduced a proof-of-concept simulator for reproducing cardiac excitation and contraction in the human heart. By utilizing human computer tomography and magnetic resonance images, the researchers successfully developed a comprehensive model of the entire heart, including all four chambers interconnected by four valves, incorporating various aspects of its functionality, such as blood flow dynamics, mechanical behavior and electrical impulses. This model integrated a human heart simulator, enabling the exploration of various clinical parameters and facilitating device design and treatment planning for cardiac diseases and dysfunctions. Another study in [45] describes the development of a DT of the liver with the aim of enhancing our understanding of liver disease and its correlation with drug toxicity, which is a leading cause of drug failures in clinical settings.

Moreover, several studies have been conducted with the aim of enhancing our comprehension or control of specific conditions or diseases and the care process [21]. The author [3] discusses the DTs design for the management of multiple sclerosis, a chronic autoimmune and degenerative disease that affects the central nervous system.

DTs have emerged as a promising tool in the field of multiple sclerosis (MS) and are particularly well-suited for MS due to the complex and heterogeneous nature of the disease, the multitude of treatment options available, and the need for comprehensive data integration and analysis. By integrating big data analysis and ML techniques, DTs can provide a comprehensive visualization of the disease progression and enable more informed therapeutic decisions (e.g., enhancing disease characterization, predicting disease course and conducting deep clinical phenotyping of individuals). Another application is trauma management, where effective management of trauma is highly critical in time-sensitive medical conditions. The author [46] introduced an initial case study that focuses on utilizing agent-based DTs for the management of severe traumas. This includes the prehospital phase, where physicians provide initial aid to patients and transfer them to the hospital emergency department, as well as the operative phase, where the trauma team provides necessary care in the hospital emergency setting. While the implementation of such systems is still in progress, a prototype has been developed to showcase the potential of this approach. In the case of elderly management, [8] introduced a framework called CloudDTH for managing the healthcare of elderly individuals. The framework specifically addresses the challenges related to real-time monitoring and accurate crisis warnings in healthcare services for elderly patients. Finally, the application of DTs extends to diabetes management as well. The author [47] introduced the DT model employed in diabetes management tracks various aspects such as nutrition, sleep patterns and changes in physical activity. DT model continuously monitors important health parameters for diabetes management including patients' blood sugar levels, liver function, weight and more. Ongoing clinical trials have indicated that providing daily precision nutrition guidance, which relies on a continuous glucose monitoring system (CGM), food intake data and ML algorithms, can offer substantial benefits to individuals diagnosed with type 2 diabetes.

4.2 Case study: Developing a DT of the eye

While significant efforts have been invested in the development of DTs for various human organs, including the heart, brain and liver, the field of ophthalmology currently lacks any such prototype or application, and this could be explained by several factors.

The main reason can be found in the peculiarity of the eye compared to other organs. For example, creating a DT of the heart can be, in certain respects, less complex than creating one for the eye. This is largely due to the differences in the scale and nature of the functional components involved in these two organs. One of the fundamental functions of the heart – the dynamics of blood flow – operates on a macroscopic level [18]. This involves the larger structures of the heart, including the chambers and valves, which are more easily accessible and observable. In contrast, the key functions of the eye involve microscopic structures that are more challenging to examine in detail and replicate digitally. Furthermore, the study and simulation of specific interventions in the heart, such as the functioning of a new valve, is inherently a macroscopic event. It does not necessitate an exploration of microscopic structures or phenomena. Therefore, the development of a DT for the heart can focus primarily on these larger, more observable elements, simplifying the task of creating a functionally representative model [6]. Additionally, the heart's relatively exposed location within the chest cavity makes it more accessible for detailed scanning and data collection, a critical step in creating a DT. This accessibility is less straightforward in the case of the eye, which is largely shielded within the orbital bones.

Overall, the eye is a composite organ, constituted of various distinct tissues, each with its own unique function and characteristics. These include the cornea, sclera, uvea and retina, with their microscopic structures. Each of these structures plays a crucial role in the overall function of the eye, and thus, any comprehensive digital representation of the eye must incorporate the intricate interactions between these tissues. AI has been instrumental in studying these separate structures (e.g., optical coherence tomography or topography), but it has yet to reach the level of sophistication required to integrate these diverse elements into a single, coherent model [48]. The vision process is a result of the harmonious functioning of all these diverse elements. As such, the development of a DT that would reproduce these intricate interactions could significantly advance research in ophthalmology with potential advantages spanning from precision medicine to education, diagnostics and medical research.

To construct more sophisticated models, it would be first necessary to generate a static reproduction of a specific eye. This preliminary step entails the integration of images derived from both the anterior and posterior segments of the eye. The primary challenge for engineers in this endeavor is the synthesis of data originating from a multitude of sources. The construction of a comprehensive static model necessitates the utilization and interpretation of raw data and images from different technologies such as slit lamp imaging, topography, optical coherence tomography of the anterior and posterior segment, confocal microscopy, gonioscopy, echography, etc. The task of harmonizing such varied and extensive data sets would be the first challenge. Further, certain regions of the eye, such as the vitreous base, the extreme periphery of the fundus or the ciliary body, are difficult to capture with the imaging technology currently available. In these instances, AI could be employed to supplement the missing data. For example, in an eye with severe myopia, it is probable that its peripheral retina would be thinner and that retinal degenerations would be more frequent. AI algorithms could be developed to acknowledge these alterations and fill in the gaps in the data accordingly [49]. Consequently, the resulting static model, enhanced by AI, could offer a reliable and comprehensive representation of the eye. Another example is the use of confocal microscopy, an imaging tool that enables high-resolution imaging of the cornea, providing in vivo images of its structure. However, performing confocal microscopy across the entirety of the cornea - due to the enormous data requirements and time needed for comprehensive image acquisition – would be a challenge. Here, again, AI algorithms employing data from a set of representative samples obtained through confocal microscopy could predict the corneal structure in areas that were not directly imaged [50]. This method, combining direct imaging with intelligent prediction, could feasibly construct a complete, high-resolution model of the cornea.

Even though the creation of a DT of the eye would initially involve the development of a static model, the ultimate goal of this research field would not be limited to just an anatomical representation; it would be a functional replica capable of imitating both structural and functional attributes specific to an individual's eye. This could, in theory, enable the DT to react to treatments and surgeries much like their physical counterparts, offering interesting opportunities for the exploration and testing of therapies and surgical procedures. One potential application could be clinical trials for rare conditions, such as Retinitis pigmentosa. The rarity of such conditions often makes it challenging to conduct traditional clinical trials due to the lack of a large patient sample or ethical issues [51]. However, with the generation of multiple DTs reproducing the specifics of these rare diseases, it would become possible to test new treatments within a digital environment. This could significantly accelerate the process of therapy development. Moreover, the DT of the eye could be instrumental in the ethical training of young doctors, providing a realistic, risk-free environment for surgical practice. Trainees could refine their skills, reduce potential complications and increase their confidence before moving on to real-life surgeries on a specific eye and predict outcomes [52]. Finally, by accurately mirroring a patient's unique eye characteristics, the DT could simulate visual experiences, effectively predicting the individual's visual acuity and field [53]. This predictive capability could have vital applications in legal and rehabilitative contexts. In legal situations where visual capability is a determinant factor, such as in disability claims or determining fitness to drive, the DT could provide a comprehensive, objective measure of a person's visual function. It would serve as a reliable tool for accurately assessing visual impairment levels and substantiating legal claims. Furthermore, in the realm of visual rehabilitation, a DT could offer indispensable insights. By simulating a patient's visual experience, clinicians could tailor rehabilitative strategies to address specific visual deficits, enhancing the effectiveness of the rehabilitation process [54]. Finally, with the assistance of AI, a DT of the eye could integrate various biomarkers of ocular diseases such as diabetic retinopathy [55]. This would allow for the prediction of disease progression and treatment outcomes in a specific eye.

Despite being in the early stages, research into developing a DT of the eye holds considerable potential. The complexity of the task, involving extensive data integration and sophisticated computational methods, is substantial. However, given the far-reaching implications for individualized diagnostics, tailored treatment strategies and surgical training, the potential benefits underscore the value of continued research in this direction.

4.3 Other applications

4.3.1 Clinical trials design

The idea of incorporating virtual patients into the clinical trial design is an evolving concept [56]. DTs have the potential to enhance randomized controlled trials (RCTs) by reducing the number of subjects required to achieve the desired statistical power [57]. Current empirical trials have limitations because they often exclude patients with comorbidities or complex treatment regimens [18]. Additionally, clinical studies face delays in the enrollment phase, and some trials fail to meet overall enrollment goals. DTs have the potential to create unlimited virtual replicas of actual patients, enabling computational treatment with a wide range of drug combinations that can serve as the control group. This approach allows for the testing of early-stage drugs on DTs of real patients, accelerating clinical research, mitigating potential risks and reducing the need for costly trials to approve new therapies. By leveraging DT technology, the impact of hazardous drugs can be minimized, while the overall process of drug development and approval can be improved [2].

In comparative clinical trials, the use of a control group can sometimes raise ethical concerns, particularly when the treatment being tested has the potential to save lives, and the standard of care or placebo is not considered effective. Ethical issues can also arise when there are significant differences in the characteristics of the treatments being compared, such as safety concerns or invasive procedures compared to noninvasive ones. DTs have the potential to address these ethical issues by replacing placebo or standard-of-care patients with virtual counterparts that simulate the evolution of health states based on patients' characteristics. By doing so, DTs can provide a representative view of how an intervention may impact the VT, effectively creating a synthetic control group [2].

While virtual clinical trials have demonstrated significant potential and offer several advantages, they have not yet reached a stage where they can completely replace human trials [56]. There are several issues to be considered including ensuring the accuracy and reliability of the VT's simulation, addressing potential biases in the data used to create the DT, and maintaining transparency in how the DT is developed and utilized. Addressing these limitations in DTs in clinical trials could pave the way toward more targeted and efficient patient trials in the development of efficient drugs and medical devices.

4.3.2 Optimization of hospital operations

Another significant application of DT technology lies in the optimization of the entire operations management of hospitals. At the healthcare facilities and at the individual department level, one approach to improving processes involves creating testable scenarios based on real-time data inputs within a DT system. These scenarios aim to enhance various aspects such as staff allocation, visitor and patient flow management, reducing waiting times, optimizing equipment and resource allocation, facilitating emergency vehicle access, and improving overall service-related operations [6, 46]. Notably, GE Healthcare and Siemens Healthineers have developed DTs specifically for hospital management optimization. DTs in digital process optimization in hospitals facilitate the optimization of digital processes by enabling predictive capabilities and capacity planning based on patient activity and demand. In this context, DTs can analyze historical data, current trends and other relevant factors to forecast patient flow and resource requirements accurately. Additionally, they allow for the execution of workflow simulations, enabling the testing and evaluation of various operational scenarios and layouts. This capability helps hospital administrators, and decision-makers assess the potential impact of different process changes or optimizations before implementing them in the physical environment [26].

5. Limitations and open issues

DT applications face several challenges and concerns that could impede the realization of their full potential. Especially, the multidisciplinary nature of designing and developing health DT systems presents a significant challenge. While collaboration across various fields of research can potentially lead to breakthroughs, it can also impede progress [1]. Indeed, DTs combine emerging technologies, such as AI, IoT, big data and XR, and each component brings its own socioethical issues and technical limitations to the implementation stage, resulting in a lack of standardization and, thus, slower outcomes [1, 58]. The absence of standardized practices impacts various aspects, including security, privacy, interactions, roles, contribution protocols, data transmission and synchronization between the VTs and PTs. The establishment of global standards would play a crucial role in accelerating the widespread adoption of DTs and making them a reality more quickly. By having universally accepted standards, organizations and industries can benefit from streamlined processes and interoperability, fostering a more efficient and effective utilization of DT technology [59]. Although DTs can show great performance in some tasks (e.g., predictive capability), DT capabilities could not be considered yet sufficient for therapy selection and preventive care [2].

5.1 Technical limitations

DTs are indeed a complex combination of emerging technologies, each with its own set of limitations. While DTs offer numerous benefits, it is important to acknowledge the challenges associated with the individual technologies involved. For instance, while the cost-effectiveness and ease of implementation of IoT devices have facilitated increased connectivity, it is important to recognize the persistent challenges associated with their use. Issues such as power outages, software errors and ongoing deployment errors continue to pose obstacles. In the healthcare domain, these challenges become particularly critical, as the disconnection of sensors that provide data to health AI algorithms can significantly impact the objective of real-time monitoring [1]. When taking into account the advancements in AI technology, the full potential of ML in healthcare is hindered by the underutilization of existing medical data, primarily due to data silos and privacy concerns that limit access to this valuable information. Without access to an ample amount of data, ML faces obstacles in effectively transitioning from research to clinical practice [28]. For instance, training an AI-based tumor detector poses challenges due to the need for a vast database covering diverse anatomies, pathologies and input data types. However, obtaining such data is difficult as health data is highly sensitive and subject to stringent regulations. Even if data anonymization is employed, it is increasingly recognized that removing identifiable information alone may not adequately protect privacy. Additionally, the process of collecting, curating and maintaining a high-quality dataset requires significant time, effort and financial investment. As a result, these data sets possess substantial business value, leading to a decreased likelihood of their free sharing. Data collectors tend to maintain strict control over the data they have gathered, retaining fine-grained ownership and access rights. Another challenge refers to VR technology, where issues related to the VR interaction design, networking optimization and optimized hardware controls need to be addressed [30].

5.1.1 Data diversity and multisourcing

The progress of AI-integrated DTs heavily depends on data fusion, which entails integrating diverse information from multiple sources. However, one of the significant challenges for human DTs is the heterogeneity and operational complexity of EHRs and healthcare information systems. For instance, dealing with different health data sources such as EHR data and imaging reports, creates inefficiency in data coding and sharing [7].

5.1.2 Data bias

A crucial concern revolves around building DT technology on data sets that contain biased patterns. Since DT technology focuses on identifying patterns, training algorithms with flawed data can amplify and perpetuate those biases. Unfortunately, many existing data sets contain biases based on factors such as race, gender or other demographics. Utilizing these data sets without appropriate correction can perpetuate and amplify these biases, leading to DTs making suboptimal or inappropriate recommendations, particularly for individuals who do not align with the "ideal" demographic profile. It is essential that DT designers address this concern to ensure that DT systems do not inadvertently reinforce existing biases present in the data, which could have negative implications for decision-making and outcomes [19].

5.1.3 Overconfidence in data and models

The consequences of bad data, flawed analysis and subsequent inaccurate representation are magnified due to the trust placed in these models. It is crucial to remain cautious about the overreliance on data and ensure robust data validation and analysis methods to mitigate the risks associated with overconfidence in the results produced by DT models [58].

5.2 Socioethical issues

5.2.1 Security and privacy

The primary socioethical risk that stands out is the violation of security and privacy. The ethical concern surrounding healthcare organizations, insurance companies or any other entities possessing a persistent and detailed record of an individual's biological, genetic, physical and lifestyle information over an extended period is a troubling issue with significant implications. The protection of DT systems from unauthorized access, misuse, modification or disclosure presents a significant challenge, similar to any other information system. Given that DT systems handle large volumes of sensitive and personal data, they become attractive targets for threat actors and cyber-attacks. Moreover, the integration of IoT devices and sensors further complicates the implementation of adequate security measures, as traditional security controls may not be well-suited for these components.

Furthermore, processing personal user data within DT systems introduces regulatory risks. Compliance with privacy regulations such as the General Data Protection Regulation (GDPR) in Europe or relevant national data protection laws becomes mandatory, which adds further complexity when designing DT systems and imposes additional challenges [18]. The future objective should be to prioritize the privacy of the data utilized in DT applications [1, 8, 58].

5.2.2 Change of structures and roles in organizations

DT stakeholders have raised concerns about the risks associated with institutional changes related to DTs, even if they may seem minor initially. One significant concern relates to the question of diagnostic responsibility once a DT becomes involved in the diagnostic process. If the real-life physician remains the primary diagnostician, what happens if they override the AI-based component of a DT? Similarly, what happens if a wrong diagnosis is made based on the data from the DT? To address these concerns, some stakeholders are now limiting the influence of DTs: while a DT can provide valuable insights into different intervention scenarios, the decision-making process remains in the hands of the human physician. These issues gain importance as computers become increasingly intelligent, while policies and regulations struggle to keep up [58].

5.2.3 Lack of trust in DT systems

There is still a prevailing lack of confidence among doctors when it comes to relying on AI algorithms and big data for decision-making in real-world problems. The primary reason behind this skepticism is the absence of clear and comprehensible explanations to support the predictions made by these systems: To what extent can we rely on the predictions made by ML models, and how accurate are these models? [59, 60]. Second, DTs are seen as flawed due to their reliance on devices to transfer data. This poses a significant gap in the reliability of DT since these devices can experience crashes or disconnections for various reasons [59]. A recent study investigating the incorporation of AI systems in hospital settings reveals that numerous physicians maintain a skeptical attitude toward AI due to the significant risks associated with possible misdiagnoses and inappropriate treatment [60]. Therefore, establishing trust and confidence in the concept of DT as a whole necessitates the establishment of standards, raising awareness and advancing technologies, all of which require significant effort and time [59].

5.2.4 Inequality and injustice in terms of accessibility to technology

The utilization of DT technology can contribute to inequality and other forms of injustice. Certainly, since it is a relatively new technology, not everyone may have access to it or be covered by health insurance that includes DT services. This can further widen existing socioeconomic disparities.

5.2.5 Human enhancement and good gene pool

The potential for predicting lifespan could be based on a combination of genetic makeup and lifestyle information found in someone's DT. By analyzing data from DTs, it may be possible to identify clusters of people with different life expectancies and distinguish those prone to leading long and healthy lives from others. This medically relevant distinction can be built upon existing statistical patterns in the population of DTs. If certain lifestyle factors associated with long life are discovered, efforts can be made to encourage more people to adopt those healthier habits using various incentives. The question arises as to whether this life extension or betterment achieved through such means should be considered therapy or enhancement [61]. While the debate regarding the distinction between therapy and enhancement is ongoing, there is currently no consensus regarding the intended purpose of DTs [62]. A clearer understanding of the differentiation between therapy and enhancement in the context of DT is likely to emerge as DT applications mature and receive broader attention and engagement from the scientific community, policymakers and regulators.

6. Conclusions

DTs in healthcare represent a promising and (to some extent) revolutionary convergence of advanced technologies such as AI, IoT, Big Data and VR to create applications that can benefit human health and health systems' efficiency. They have transformative potential across various dimensions of healthcare, from diagnostics and therapy planning to medical education and clinical trial design. Their inherent capabilities to deliver personalized, predictive and dynamic models of individual patients could vastly improve health outcomes, and their ability to reproduce complex systems and to exploit the available informative set bears the potential to create much-needed efficiencies for both healthcare providers and technology producers. DTs provide exciting possibilities in diagnostics and therapy planning, through which physicians can visualize and understand a patient's health state in real time, enabling personalized and timely interventions. Medical education stands to benefit significantly from DTs, providing immersive, realistic training without the risks associated with training on real patients. Furthermore, DTs can contribute substantially to clinical trials, addressing issues such as ethical concerns, participant selection and trial design. They also hold the promise to optimize hospital operations and service delivery, thereby further enhancing the quality of healthcare and optimizing the use of healthcare resources.

However, DTs are not devoid of challenges. On the technical side, the integration of multiple technologies, most of which were not initially developed to work in integration, and the need for synchronized operation presents a formidable challenge. The quality of data that forms the basis of DTs, including potential bias, overconfidence in models, and diversity of data sources, are just some of the current limitations of DTs that need to be critically assessed to ensure accuracy and reliability. Moreover, the socioethical implications of DTs are complex. Issues surrounding security and privacy, changes in structures and roles within healthcare institutions and in clinical decision-making, trust in DT systems and accessibility inequalities necessitate rigorous examination and appropriate policy responses. These concerns become especially pertinent given the sensitive and personal nature of the data involved in DTs, which makes them attractive targets for cyber threats and raises ethical considerations. The advent of DTs also brings up intriguing discussions on the subject of human health enhancement, as DTs may eventually facilitate personalized health optimization strategies. However, this must be cautiously approached, ensuring it does not foster (or even boost) inequality or violate ethical principles.

We believe that healthcare institutions must proactively navigate these challenges, considering both the technical complexities and the broader socioethical implications. Developing global standards for DT technology, fostering trust and understanding among medical practitioners, and ensuring adherence to security and privacy regulations will be essential steps toward realizing the full potential of DTs. Future efforts should be directed toward advancing the technology, promoting its understanding among key stakeholders, and establishing robust policies that strike a balance between leveraging the immense potential of DTs and addressing the associated challenges. A holistic and ad-hoc approach that integrates technical advancement, regulatory compliance and ethical considerations will be key to unlocking the vast potential of DTs for global healthcare improvements.

Acknowledgements

This research was conducted by the SDA Bocconi's LIFT Lab in 2021, and at that time, the funding members were: AB MEDICA SPA, BRACCO IMAGING SPA, FOND. E.A. FIERA INTERN. MILANO, INTESA SANPAOLO FORMAZIONE SPA, INTESA SANPAOLO INNOVATION CENTER SPA, KIROMIC BIOPHARMA INC., MEDTRONIC ITALIA SPA, WENNOVIA SRL.

Conflict of interest

The authors declare no conflict of interest.

Author details

Patrizio Armeni[1*], Irem Polat[2], Leonardo Maria De Rossi[3], Lorenzo Diaferia[3], Giacomo Visioli[4], Severino Meregalli[3] and Anna Gatti[2]

1 LIFT Lab and CERGAS, Claudio Demattè Research Division and GHNP Division, SDA Bocconi School of Management, Milano, Italy

2 LIFT Lab, Claudio Demattè Research Division and GHNP Division, SDA Bocconi School of Management, Milano, Italy

3 LIFT Lab and DEVO Lab, Claudio Demattè Research Division and GHNP Division, SDA Bocconi School of Management, Milano, Italy

4 Sense Organs Department, Sapienza – University of Rome, Rome, Italy

*Address all correspondence to: patrizio.armeni@unibocconi.it

References

[1] Fuller A, Fan Z, Day C, Barlow C. Digital twin: Enabling technologies, challenges and open research. IEEE Access. 2020;**8**:108952-108971. DOI: 10.1109/ACCESS.2020.2998358

[2] Armeni P, Polat I, De Rossi LM, Diaferia L, Meregalli S, Gatti A. Digital twins in healthcare: Is it the beginning of a new era of evidence-based medicine? A critical review. Journal of Personalized Medicine. 2022;**12**(8):1255. DOI: 10.3390/jpm12081255

[3] Voigt I, Inojosa H, Dillenseger A, Haase R, Akgün K, Ziemssen T. Digital twins for multiple sclerosis. Frontiers in Immunology. 2021;**12**:669811. DOI: 10.3389/fimmu.2021.669811

[4] Barricelli BR, Casiraghi E, Gliozzo J, Petrini A, Valtolina S. Human digital twin for fitness management. IEEE Access. 2020;**8**:26637-26664. DOI: 10.1109/ACCESS.2020.2971576

[5] Lu Y, Liu C, Kevin I, Wang K, Huang H, Xu X. Digital Twin-driven smart manufacturing: Connotation, reference model, applications and research issues. Robotics and Computer-Integrated Manufacturing. 2020;**61**:101837. DOI: 10.1016/j.rcim.2019.101837

[6] Coorey G, Figtree GA, Fletcher DF, Snelson VJ, Vernon ST, Winlaw D, et al. The health digital twin to tackle cardiovascular disease—A review of an emerging interdisciplinary field. NPJ Digital Medicine. 2022;**5**(1):126. DOI: 10.1038/s41746-022-00640-7

[7] Venkatesh KP, Raza MM, Kvedar JC. Health digital twins as tools for precision medicine: Considerations for computation, implementation, and regulation. npj Digital Medicine. 2022;**5**(1):150. DOI: 10.1038/s41746-022-00694-7

[8] Liu Y, Zhang L, Yang Y, Zhou L, Ren L, Wang F, et al. A novel cloud-based framework for the elderly healthcare services using digital twin. IEEE Access. 2019;**7**:49088-49101. DOI: 10.1109/ACCESS.2019.2909828

[9] Okegbile SD, Cai J, Yi C, Niyato D. Human digital twin for personalized healthcare: Vision, architecture and future directions. IEEE Network. 2022;**2022**:1-7. DOI: 10.1109/MNET.118.2200071

[10] Khan S, Arslan T, Ratnarajah T. Digital twin perspective of fourth industrial and healthcare revolution. IEEE Access. 2022;**10**:25732-25754. DOI: 10.1109/ACCESS.2022.3156062

[11] Topol EJ. High-performance medicine: The convergence of human and artificial intelligence. Nature Medicine. 2019;**25**(1):44-56. DOI: 10.1038/s41591-018-0300-7

[12] Tao F, Sui F, Liu A, Qi Q, Zhang M, Song B, et al. Digital twin-driven product design framework. International Journal of Production Research. 2019;**57**(12):3935-3953. DOI: 10.1080/00207543.2018.1443229

[13] Acosta JN, Falcone GJ, Rajpurkar P, Topol EJ. Multimodal biomedical AI. Nature Medicine. 2022;**28**(9):1773-1784. DOI: 10.1038/s41591-022-01981-2

[14] Dhruva SS, Ross JS, Akar JG, Caldwell B, Childers K, Chow W, et al. Aggregating multiple real-world data sources using a patient-centered health-data-sharing platform. NPJ Digital

Medicine. 2020;**3**(1):60. DOI: 10.1038/s41746-020-0265-z

[15] Raden N. Digital Twins for Personalized Medicine-a Critical Assessment. Diginomica [Internet]. 2020. Available from: https://diginomica.com/digital-twins-personalized-medicine-critical-assessment [Accessed: April 23, 2023]

[16] Marsch LA. Digital health data-driven approaches to understand human behavior. Neuropsychopharmacology. 2021;**46**(1):191-196. DOI: 10.1038/s41386-020-0761-5

[17] Coravos A, Khozin S, Mandl KD. Developing and adopting safe and effective digital biomarkers to improve patient outcomes. NPJ Digital Medicine. 2019;**2**(1):14. DOI: 10.1038/s41746-019-0090-4

[18] Corral-Acero J, Margara F, Marciniak M, Rodero C, Loncaric F, Feng Y, et al. The 'Digital Twin' to enable the vision of precision cardiology. European Heart Journal. 2020;**41**(48):4556-4564. DOI: 10.1093/eurheartj/ehaa159

[19] Schwartz SM, Wildenhaus K, Bucher A, Byrd B. Digital twins and the emerging science of self: Implications for digital health experience design and "small" data. Frontiers in Computer Science. 2020;**2**:31. DOI: 10.3389/fcomp.2020.00031

[20] Rasheed A, San O, Kvamsdal T. Digital twin: Values, challenges and enablers from a modeling perspective. IEEE Access. 2020;**8**:21980-22012. DOI: 10.1109/ACCESS.2020.2970143

[21] Kamel Boulos MN, Zhang P. Digital twins: From personalised medicine to precision public health. Journal of Personalized Medicine. 2021;**11**(8):745. DOI: 10.3390/jpm11080745

[22] Ahmadi-Assalemi G, Al-Khateeb H, Maple C, Epiphaniou G, Alhaboby ZA, Alkaabi S, et al. Digital twins for precision healthcare. In: Jahankhani H, Kendzierskyj S, Chelvachandran N, Ibarra J, editors. Cyber Defence in the Age of AI, Smart Societies and Augmented Humanity. Cham: Springer; 2020. pp. 133-158. DOI: 10.1007/978-3-030-35746-7_8

[23] Zhang Z, Wen F, Sun Z, Guo X, He T, Lee C. Artificial intelligence-enabled sensing technologies in the 5G/internet of things era: From virtual reality/augmented reality to the digital twin. Advanced Intelligent Systems. 2022;**4**(7):2100228. DOI: 10.1002/aisy.202100228

[24] Nguyen HX, Trestian R, To D, Tatipamula M. Digital twin for 5G and beyond. IEEE Communications Magazine. 2021;**59**(2):10-15. DOI: 10.1109/MCOM.001.2000343

[25] Erol T, Mendi AF, Doğan D. The digital twin revolution in healthcare. In: 2020 4th International Symposium on Multidisciplinary Studies and Innovative Technologies (ISMSIT); 22-24 October 2020. Istanbul, Turkey: IEEE; 2020. pp. 1-7. DOI: 10.1109/ISMSIT50672.2020.9255249

[26] Barricelli BR, Casiraghi E, Fogli D. A survey on digital twin: Definitions, characteristics, applications, and design implications. IEEE Access. 2019;**7**:167653-167671. DOI: 10.1109/ACCESS.2019.2953499

[27] Nagaraj D, Khandelwal P, Steyaert S, Gevaert O. Augmenting digital twins with federated learning in medicine. The Lancet Digital Health. 2023;**5**(5):E251-E253. DOI: 10.1016/S2589-7500(23)00044-4

[28] Rieke N, Hancox J, Li W, Milletari F, Roth HR, Albarqouni S, et al. The future

of digital health with federated learning. NPJ Digital Medicine. 2020;**3**(1):119. DOI: 10.1038/s41746-020-00323-1

[29] Künz A, Rosmann S, Loria E, Pirker J. The potential of augmented reality for digital twins: A literature review. In: 2022 IEEE Conference on Virtual Reality and 3D User Interfaces (VR); 12-16 March 2022. Christchurch, New Zealand: IEEE; 2022. pp. 389-398. DOI: 10.1109/VR51125.2022.00058

[30] Pirker J, Loria E, Safikhani S, Künz A, Rosmann S. Immersive virtual reality for virtual and digital twins: A literature review to identify state of the art and perspectives. In: 2022 IEEE Conference on Virtual Reality and 3D User Interfaces Abstracts and Workshops (VRW); 12-16 March 2022. Christchurch, New Zealand: IEEE; 2022. pp. 114-115. DOI: 10.1109/VRW55335.2022.00035

[31] Laaki H, Miche Y, Tammi K. Prototyping a digital twin for real time remote control over mobile networks: Application of remote surgery. IEEE Access. 2019;7:20325-20336. DOI: 10.1109/ACCESS.2019.2897018

[32] Zheng Y, Lu R, Guan Y, Zhang S, Shao J. Towards private similarity query based healthcare monitoring over digital twin cloud platform. In: 2021 IEEE/ACM 29th International Symposium on Quality of Service (IWQOS); 25-28 June 2021. Tokyo, Japan: IEEE; 2021. pp. 1-10. DOI: 10.1109/IWQOS52092.2021.9521351

[33] Alazab M, Khan LU, Koppu S, Ramu SP, Iyapparaja M, Boobalan P, et al. Digital twins for healthcare 4.0-recent advances, architecture, and open challenges. IEEE Consumer Electronics Magazine. 2022;**2022**:1-8. DOI: 10.1109/MCE.2022.3208986

[34] Martinez-Velazquez R, Gamez R, el Saddik A. Cardio Twin: A Digital Twin of the human heart running on the edge. In: 2019 IEEE International Symposium on Medical Measurements and Applications (MeMeA); 26-28 June 2019. Istanbul, Turkey: IEEE; 2019. pp. 1-6. DOI: 10.1109/MeMeA.2019.8802162

[35] Okegbile SD, Cai J. Edge-assisted human-to-virtual twin connectivity scheme for human digital twin frameworks. In: 2022 IEEE 95th Vehicular Technology Conference:(VTC2022-Spring); 19-22 June 2022. Helsinki, Finland: IEEE; 2022. pp. 1-6. DOI: 10.1109/VTC2022-Spring54318.2022.9860619

[36] Akash SS, Ferdous MS. A Blockchain based system for healthcare digital twin. IEEE Access. 2022;**10**:50523-50547. DOI: 10.1109/ACCESS.2022.3173617

[37] Yaqoob I, Salah K, Uddin M, Jayaraman R, Omar M, Imran M. Blockchain for digital twins: Recent advances and future research challenges. IEEE Network. 2020;**34**(5):290-298. DOI: 10.1109/MNET.001.1900661

[38] Raj P. Empowering digital twins with blockchain. In: Aggarwal S, Kumar N, Raj P, editors. Advances in Computers. Amsterdam: Elsevier; 2021. pp. 267-283. DOI: 10.1016/bs.adcom.2020.08.013

[39] FDA (US Food and Drug Administration). Precision Medicine [Internet]. 2018. Available from: https://www.fda.gov/medical-devices/in-vitro-diagnostics/precision-medicine [Accessed: April 29, 2023]

[40] Joyner MJ, Paneth N. Promises, promises, and precision medicine. The Journal of Clinical Investigation. 2019;**129**(3):946-948. DOI: 10.1172/JCI126119

[41] Denny JC, Collins FS. Precision medicine in 2030—Seven ways

to transform healthcare. Cell. 2021;**184**(6):1415-1419. DOI: 10.1016/j.cell.2021.01.015

[42] Björnsson B, Borrebaeck C, Elander N, Gasslander T, Gawel DR, Gustafsson M, et al. Digital twins to personalize medicine. Genome Medicine. 2020;**12**:1-4. DOI: 10.1186/s13073-019-0701-3

[43] Baillargeon B, Rebelo N, Fox DD, Taylor RL, Kuhl E. The living heart project: A robust and integrative simulator for human heart function. European Journal of Mechanics-A/Solids. 2014;**48**:38-47. DOI: 10.1016/j.euromechsol.2014.04.001

[44] Jirsa V, Wang H, Triebkorn P, Hashemi M, Jha J, Gonzalez-Martinez J, et al. Personalised virtual brain models in epilepsy. The Lancet Neurology. 2023;**22**(5):443-454. DOI: 10.1016/S1474-4422(23)00008-X

[45] Subramanian K. Digital twin for drug discovery and development—The virtual liver. Journal of the Indian Institute of Science. 2020;**100**(4):653-662. DOI: 10.1007/s41745-020-00185-2

[46] Croatti A, Gabellini M, Montagna S, Ricci A. On the integration of agents and digital twins in healthcare. Journal of Medical Systems. 2020;**44**:1-8. DOI: 10.1007/s10916-020-01623-5

[47] Shamanna P, Saboo B, Damodharan S, Mohammed J, Mohamed M, Poon T, et al. Reducing HbA1c in type 2 diabetes using digital twin technology-enabled precision nutrition: A retrospective analysis. Diabetes Therapy. 2020;**11**:2703-2714. DOI: 10.1007/s13300-020-00931-w

[48] Honavar SG. Artificial intelligence in ophthalmology-machines think! Indian Journal of Ophthalmology. 2022;**70**(4):1075-1079. DOI: 10.4103/ijo.IJO_644_22

[49] Jin K, Ye J. Artificial intelligence and deep learning in ophthalmology: Current status and future perspectives. Advances in Ophthalmology Practice and Research. 2022;**2**(3):100078. DOI: 10.1016/j.aopr.2022.100078

[50] Alam U, Anson M, Meng Y, Preston F, Kirthi V, Jackson TL, et al. Artificial intelligence and corneal confocal microscopy: The start of a beautiful relationship. Journal of Clinical Medicine. 2022;**11**(20):6199. DOI: 10.3390/jcm11206199

[51] Singh MS. Eyeing a role for artificial intelligence in clinical trials. JAMA Ophthalmology. 2023;**141**(4):313-314. DOI: 10.1001/jamaophthalmol.2022.6542

[52] Ferris JD, Donachie PH, Johnston RL, Barnes B, Olaitan M, Sparrow JM. Royal College of ophthalmologists' National Ophthalmology Database study of cataract surgery: Report 6. The impact of EyeSi virtual reality training on complications rates of cataract surgery performed by first and second year trainees. British Journal of Ophthalmology. 2020;**104**(3):324-329

[53] Lam AK, To E, Weinreb RN, Yu M, Mak H, Lai G, et al. Use of virtual reality simulation to identify vision-related disability in patients with glaucoma. JAMA Ophthalmology. 2020;**138**(5):490-498. DOI: 10.1001/jamaophthalmol.2020.0392

[54] Porter LF, Black GC. Personalized ophthalmology. Clinical Genetics. 2014;**86**(1):1-11. DOI: 10.1111/cge.12389

[55] Visioli G, Alisi L, Mastrogiuseppe E, Albanese GM, Romano E, Iannetti L, et al. OCT biomarkers as predictors of visual improvement in diabetic macular

edema eyes receiving dexamethasone implants. International Journal of Retina and Vitreous. 2023;**9**(1):1-10. DOI: 10.1186/s40942-023-00473-w

[56] Abadi E, Segars WP, Tsui BM, Kinahan PE, Bottenus N, Frangi AF, et al. Virtual clinical trials in medical imaging: A review. Journal of Medical Imaging. 2020;**7**(4):042805. DOI: 10.1117/1.JMI.7.4.042805

[57] Walsh JR, Roumpanis S, Bertolini D, Delmar P. Evaluating digital twins for alzheimer's disease using data from a completed Phase 2 clinical trial. Alzheimer's & Dementia. 2022;**18**:e065386. DOI: 10.1002/alz.065386

[58] Popa EO, van Hilten M, Oosterkamp E, Bogaardt MJ. The use of digital twins in healthcare: Socio-ethical benefits and socio-ethical risks. Life Sciences, Society and Policy. 2021;**17**(1):1-25. DOI: 10.1186/s40504-021-00113-x

[59] Elayan H, Aloqaily M, Guizani M. Digital twin for intelligent context-aware IoT healthcare systems. IEEE Internet of Things Journal. 2021;**8**(23):16749-16757. DOI: 10.1109/JIOT.2021.3051158

[60] Reis L, Maier C, Mattke J, Creutzenberg M, Weitzel T. Addressing user resistance would have prevented a healthcare AI project failure. MIS Quarterly. 2020;**19**(4):279-296. Available at: https://aisel.aisnet.org/misqe/vol19/iss4/8

[61] Bruynseels K, Santoni de Sio F, Van den Hoven J. Digital twins in health care: Ethical implications of an emerging engineering paradigm. Frontiers in Genetics. 2018;**9**:31. DOI: 10.3389/fgene.2018.00031

[62] Hofmann B. Limits to human enhancement: Nature, disease, therapy or betterment? BMC Medical Ethics. 2017;**18**(1):1-11. DOI: 10.1186/s12910-017-0215-8

Chapter 5

Perspective Chapter: Digital Twin Applied in the Brazilian Energy Sector

Eldrey Seolin Galindo and Urbano Chagas

Abstract

This chapter explores the applications of Digital Twin (DT) technology in the Brazilian energy sector and its impact on businesses and society. It highlights how DT applications have contributed to cost reduction, human error mitigation, operational optimization, and technical failure prediction. The chapter also discusses the implementation process and the requirements for developing these systems. Additionally, it explores the potential of leveraging Artificial Intelligence for decision-making support, utilizing Big Data processes to enhance various areas, and employing User Experience (UX) techniques to streamline outdated processes, through the examination of real projects in the wind power monitoring, transmission towers, and data-saving equipment domains, addressing the challenges faced and the benefits derived from its implementation.

Keywords: digital twin applications, Brazilian energy sector, optimization, failure prediction, development and implementation, artificial intelligence, big data

1. Introduction

To understand the challenges of implementing the Digital Twin (DT) applications in the Brazilian electricity sector, we need some context about the country and it is a complex infrastructure of energy; for this, **Table 1** compares the five biggest countries in terms of territory, where Brazil is the fifth with 8.515.770 km^2 two times smaller than Russia the biggest country. However, the Brazilian population is third, behind China and the United States, and the ninth economy in the world [1].

The Brazilian electricity sector is composed of four segments—Generation, Transmission, Distribution, and Commercialization—both public and private companies are responsible for maintaining the infrastructure and local operations under the guidance of the National System Operator (ONS). The ONS continuously monitors real-time information about the energy infrastructure, which comprises 23,441 operational power generation units distributed across the country. **Figure 1** provides a visual representation of the distribution of these power generation units, revealing the complex structure. **Table 2** gives us an overview of the distribution of energy sources used by the five biggest countries. The distribution of electricity generation sources amounts to these countries are similar with one main source upper than

Data sources	1°	2°	3°	4°	5°
	Russia	**Canada**	**United States**	**China**	**Brazil**
Total km^2	17.098.242	9.984.670	9.826.675	9.596.960	8.515.770
Population[a]	141.698.923	38.516.736	339.665.118	1.413.142.846	218.689.757
Generating capacity[b]	276.463 kW	153.251 kW	1.143.266 kW	2.217.925 kW	195.037 kW
Gross domestic product[c]	$4.078	$1.832	$21.132	$24.861	$3.128

[a]Values estimated for 2023.[b]Values in million kW and estimated for 2020.[c]Values in trillions quoted in 2017 dollars and estimated for 2021.

Table 1.
Comparison between the five largest countries by territory in the world according to The World Factbook by CIA [1].

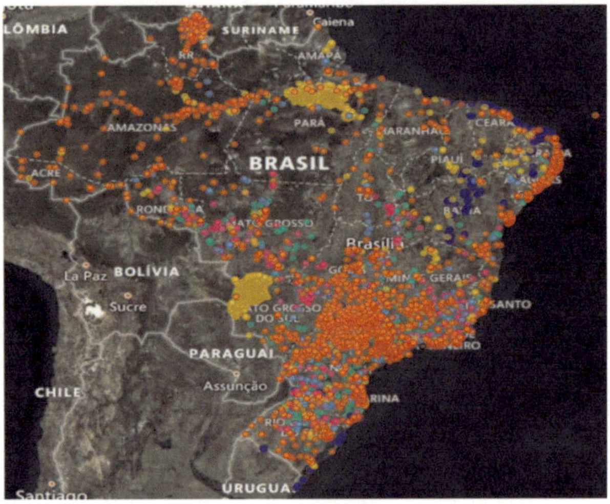

Figure 1.
Print screen of the public data system of ANEEL showing the active generating units distributed in the Brazilian territory [2].

Generation sources	1°	2°	3°	4°	5°
	Russia (%)	**Canada (%)**	**United States (%)**	**China (%)**	**Brazil (%)**
Fossil fuels	59.4	16.5	59.9	66	11.8
Nuclear	21	14.7	19.5	4.8	2.3
Solar	0.2	0.7	3.2	3.5	1.7
Wind	0	5.7	8.3	6.2	9.2
Hydroelectricity	19.1	60.8	7	17.8	65.8
Geothermal	0	0	0.4	0	0
Biomass and Waste	0.3	1.6	1.7	1.6	9.2

Table 2.
Comparative analysis of the electricity generation sources of the top five largest countries in terms of land area worldwide according to The World Factbook by CIA [1] (Estimated data for 2020).

Generation sources	Brazil	
	2020[a] (%)	2023[b] (%)
Fossil fuels	11.8	15.78
Nuclear	2.3	1.03
Solar	**1.7**	**4.47**
Wind	**9.2**	**13.39**
Hydroelectricity	65.8	56.55
Biomass and Waste	9.2	8.79

[a]*The World Factbook by CIA data [1].*[b]*National Agency of Electricity (ANEEL) data [2].*

Table 3.
Comparative analysis of the Brazilian electricity generation sources between 2020 and 2023, where in bold there are solar and wind energy sources that have a considered increase during these 3 years.

59% of generation, for Brazil and Canada the main source is hydroelectricity, and for others, fossil fuels are the main.

In addition to other emerging countries, Brazil faces significant challenges in its electricity sector. In recent years, the Brazilian government and public agencies have been actively promoting the modernization of the energy sector on two fronts, the search for renewable and clean energies and technological modernization [3]. Comparing the 2020 CIA data with the most recent 2023 ANEEL data, **Table 3**, we note that solar and wind energy sources have increased during these 3 years. Technological modernization was through the adoption of advanced technologies such as Artificial Intelligence, Big Data, Internet of Things (IoT), and DT.

These technologies aim to enhance the availability and predictability of the national electrical network. While the first DT projects in Brazil officially began in 2020, initiatives incorporating the concept of DT and utilizing data simulation, virtual monitoring, failure prediction, and power generation forecasting have been underway since 2009, as Brasilian law 13.755/Dez2018 better known as" ROTA2030" which encourages automobile modernization and increases the number of electric cars [4], and various research and development projects that the public data from the National Agency of Electricity (ANEEL) provides into this progressive development. In 2007, the Ministry of Mines and Energy (MME) published the National Energy Plan (PDE) a study and planning of energy evolution until 2030; in this study, MME estimates an annual average investment in the sector of 11.4 billion dollars [5]. According to the research by Wanasinghe et al. [6] about the trends, opportunities, and challenges of the oil and gas industry, among the 199 articles investigated between January 2003 and April 2020, Brazil appears with approximately 15 articles, occupying the fourth position behind the United States (59 articles), Norway (22 articles), and United Kingdom (16 articles).

Next, we will explore three real projects that we worked on and that will apply or have started to apply DT in this sector. We will not focus on the more technical part or the algorithms used but on the challenges encountered at the beginning of each project and mainly on those encountered by them.

2. Monitoring wind power

Wind power has always played a prominent role in pursuing clean energy. From propelling caravels and mills to driving turbines, humanity has harnessed its power for many years. However, for a long time, controlling or predicting the intensity of the wind was a challenge, making it difficult to forecast the amount of energy that could be generated. Although improvements in weather forecasts and satellite imagery have helped mitigate this issue, a definitive solution remained elusive.

In 2017, a wind farm company approached us with a request to develop a system capable of predicting power generation in their turbines and supporting maintenance decisions by identifying potential failures in advance. We began by asking fundamental questions: How do wind turbines work? And what data do we need to predict their performance?

At a basic level, the wind turbine shown in **Figure 2** consists of three main components: the blades, which harness the force of the wind and convert it into circular motion in the turbine; the nacelle, which houses the turbine's motor and the necessary equipment to convert the circular motion into electrical energy; and the tower, which provides structural support to the entire system. The positioning of the wind turbine plays a crucial role in maximizing power generation, as it constantly adjusts the direction of the nacelle and the angle of the blades to optimize energy capture. Now, we can find the first data necessary to predict their performance, the angle blades, nacelle direction, wind velocity, and the amount of energy generated. Finally, when we study the turbine sensors deeper, we find that about 10 variables were used for turbine control and other variables linked with weather and generation system.

Differently than what we initially thought, after we answered our fundamental questions and identified all variables we need to predict and improve the power generation, the following steps were not developing a machine learning algorithm or a simulation system, the project started with the aggregation of data and creation of a historical dataset. In the process of digital transformation, it is common to find companies at distinct digital levels, and in the electrical sector that is no different, many companies have their equipment with many sensors and exporting data all time, but these data are not processed or saved for a long time and the history is lost. The DT is a big concept formed of three basic parts: monitoring, simulation, and prediction, the first part is commonly implemented in this sector since many laws oblige these processes. The second and third parts are a challenge. Give me leave to do a reflection, if you are an owner of an energy company and you will buy some equipment for a new wind farm, you will search for the most resistant equipment that you can, equipment that does not break; after all, a broken turbine does not generate energy and this is loss

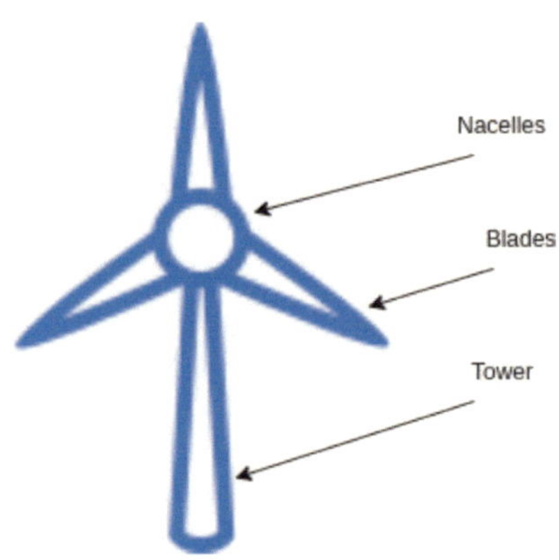

Figure 2.
Simple view of wind turbine showing the three main components (Nacelles, Blades, and Tower) [Authors].

money, to help in your quest; your energy supply contracts usually have high fines for non-supply. The reflection here is the equipment is made to not break its lifespan is very long close to 20 years, and for an ideal prediction model or simulation system, you will need to preserve decades of data at the cost of a lot of space. We have not considered on this scene that you save the correct information, the problems that occurred, the maintenance, and the manual optimizations, which are valuable targets.

For this project, we need to create historical data with some months using only the sensor data per minute, and this information was enough for us can predict failures and create alerts on an internal system to help operators in maintaining. In real operation, the application helped to reduce stops to maintaining and the generation continual. Other projects for prediction and optimization and simulation system using weather data were developed by the client.

In this case, the company had the equipment with all the necessary sensors and the know-how to operate and optimize it but had insufficient and unstructured data for a simulation or a prediction of power generated, but when the equipment does not have a sensor, it is impossible to have.

3. The challenge of transmission towers

For the energy generated in a wind farm to turn on the light in our house, such energy needs to be transmitted through power lines. At first glance, this process is not complex. A person can use an extension cord to plug in a lawnmower or a lamp in the garden every weekend. New houses are integrated into the energy distribution system all the time, and there are power poles all over the city or along the roads. However, when there is a need to connect cities, big factories, states, and in extreme cases, countries, this infrastructure becomes more complex. The challenge of transmitting energy starts with the generation of energy. The energy generated in the generation unit cannot be stored or is very expensive to do so. In summary, all the energy generated needs to be distributed in real time for consumption, and, if the energy is not used, it is lost. Another obstacle is the distance, except for the small units or home generation, either the main units are built far or their capability to make energy is the biggest that the closed cities need, and the extra energy is distributed to other places. Additionally, energy is transmitted at high voltage to overcome great distances. This is the main distinction between the energy in the city and the energy for the city. Although high voltage is used for power transmission, it is not used for distribution. According to the norms of the International Electrotechnical Commission [7], energy voltage can be divided into three classes. Low voltage is used for consumption and is normally between 110 and 230 V. Medium voltage, between 1000 and 36,000 V, is used for distribution, while high voltage, above 36,000 V, transmission. In Brazil, there are power lines between 13.8 kV (13,800 V) and 800 kV (800,000 V) with the longest power line covering a distance of 2518 km [8].

Due to the high voltages and the high risk of accidents, there are several safety regulations for the equipment involved in the transmission and for the surroundings of the power lines. Some of these lines pass through forest areas, flooded areas, rivers, and mountains, which makes access for maintenance more difficult, and other lines pass through urban and rural areas, close to people, air traffic, and highways. The maintenance of these lines is a challenge, and companies need to use two or three techniques to inspect the lines and identify operational or security risks. One of these techniques involves flying over the lines with a helicopter, which is very specific and requires special training and equipment. This type of flight is expensive and risky due to the high voltage and the risk to the pilot and operators involved in maintenance. Another technique involves a team climbing the towers to inspect and take pictures, but when the lines are live, the people cannot climb up high and have a direct view of the failures.

For the new flow of data and the new interaction with maintenance, it was necessary to use user experience techniques to identify the best interactions. This research was focused not only on the interface and human interactions but also on data and the future implementation of Artificial Intelligence. Some companies still do not give due importance to the usability of systems or procedures. When we talk about new procedures involving data, Artificial Intelligence, and DT, these human interactions need to be even more important. When we talked about this project with drones, we had a completely different process than the one used by the client company, which in itself was a challenge. However, our goal was to go beyond creating a new process and to involve the creation of a digital model of high-voltage transmission lines and towers.

The designed process was divided into four stages:

1. A drone flew over a few kilometers of a line, taking pictures of towers, cables, and surroundings;

2. The images were automatically sent to the center, where they were stored and cataloged;

3. Specialized inspectors tagged the images and identified failures or points of attention, triggering the maintenance teams according to the need;

4. The images and information were stored, creating a database of photos and transmission line information.

As a result, a database for each tower was created, and for the first time, the company had a history of failures with degradation levels. After some tests, Artificial Intelligence will start to identify and classify the image degradation to assist the specialists, and in the future, the towers will have their digital versions, helping to predict failures or maintenance.

The design process used to map the stakeholders and discover user interactions revealed a new sub-function for the system that was difficult to envision. The system was excellent for training new specialists to find failures, and it is possible to find inspected equipment and request a person in training to reinspect it and compare the results. In addition, some inspections are complex or require more care, and the system was able to allow double inspections, which validate the inspection done and increase the quality.

4. Data save equipment

Energy substations play a fundamental role in the energy system as they are responsible for voltage conversion and the delivery of energy for distribution. It is within the substation that the ONS (National System Operator) controls the flow of energy, defining routes and sequences of actions. For example, during a water crisis when thermoelectric power plants are used to supply the system, the ONS may request certain substations to connect to the power lines of these plants and integrate this energy with other energy sources. Another example is when a substation or a power line experiences issues and requires maintenance, leading to a disconnection of a part of the system. Substations have different equipment and specific layouts tailored to their activities. The structure for transforming high voltage to low voltage differs from the structure used to transmit produced energy. A typical transforming substation consists of main equipment such as circuit breakers, transformers, capacitor banks, and reactors. These kinds of equipment are organized and integrated into a circuit based on their function or load. Security equipment such as fuses and special circuit breakers are also present, and all devices are monitored by sensors that measure temperatures, voltages, amperage, and other data.

In 2020, we initiated a project focused on predicting equipment failures in substations and integrating this information with the internal system. All equipment is monitored, and all substations use a system with software and hardware elements called Supervisory Control and Data Acquisition (SCADA), which allows for the control of industrial processes, data monitoring, gathering and processing, and event logging [9].

Basically, an operation in a substation involves the operator monitoring raw data on SCADA and opening a Service Order (S.O.) if any parameter is non-compliant. Additionally, the operator frequently performs tests using a test device to check the integration of equipment. Early studies revealed a configuration error in the protective system, which was triggered when the temperature reached 51°C. However, this equipment functions normally until 65°C. This information raised curiosity as to why some alarms were activated and yet no action was taken. The answer was simple: human error. The engineer received the request regarding false alarms but found nothing unusual and ignored the request. Subsequently, the operator also started to ignore the alarms. Although this human failure was not serious, it highlighted a problem, which is a classic problem for data scientists. Humans struggle to process large amounts of data simultaneously and make correct decisions based on numbers. Machines, on the other hand, are better at processing such data and supporting human decisions. The algorithms were able to identify configuration failures and provide engineers with the necessary data to make more accurate decisions and precise configurations [10].

An overview of the data flow from SCADA to the control system is shown in **Figure 3**. However, the most important part, particularly for us, is the predictive modeling step, this is the very point where the DT occurs. To predict failures, the model establishes a normal state for the equipment and compares it with the current state. Based on certain parameters and time series data, the model can determine whether the current state is normal or abnormal. By simulating a future state, the model can precisely estimate how long it will take for the equipment to transit from a normal state to an abnormal state and why [10]. With this information, the engineer can assess the risk level and anticipate potential issues, whether they are serious problems with the equipment or secondary problems such as a sensor malfunction or communication failure.

Simulating the equipment and predicting abnormal states has brought many benefits to the company. The project's results have had an immediate impact on detecting system adjustments and promptly identifying abnormalities, thus preventing equipment failures. Some benefits of the project, however, are not directly related to it. With simulated predictions, it is possible to anticipate more serious failures and optimize human and material resources. For instance, during application tests, the model predicted a failure in temperature-related equipment. When the engineer received this information and investigated, they discovered that a communication

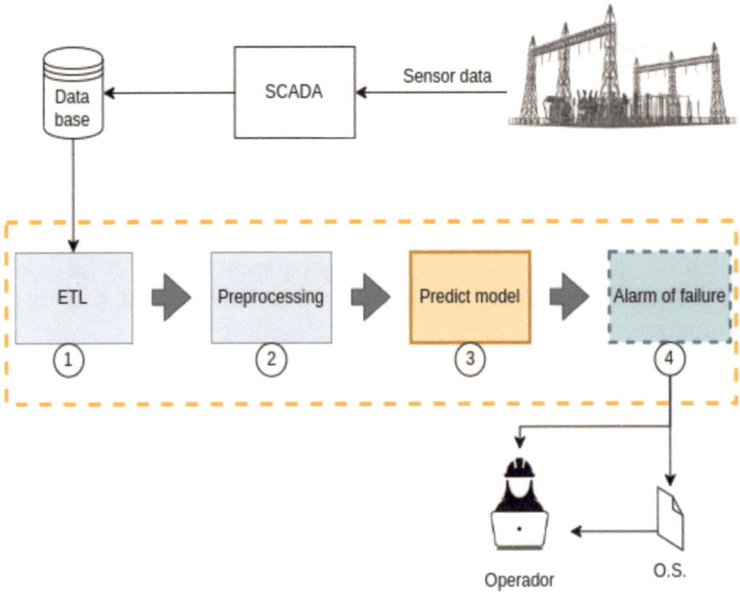

Figure 3.
General application flow diagram published on the article data mining applied to abnormality prediction in electrical substation transformers [10].

component of the sensor connected to the SCADA was starting to fail and would soon cease functioning. This raised an alert since the equipment would no longer be properly monitored, and replacing this component would take several months. By anticipating the order of the component, the time during which the equipment remained unprotected was reduced, as well as the time required for an operator to closely monitor the equipment. This resulted in significant savings in the operation of the substation.

5. Discussion

If you try to develop a DT for a company independent of the area, there is a high probability that you understand that things are not easy as they appear to be at first sight. When a client requests a project for a DT, normally the phases "the data lake was done," "the interface is simple," or "it is just to integrate with the actual systems" hide great challenges. A system that interacts with others has a clear interface with simple interaction and a correct data flow is not created only by data scientists, data engineers, and developers; in general, Brazilian energy companies does a modest investment in usability and graphical interfaces, preferring cheaper and less user-friendly projects. On the other hand, Camara et al. [11] expose some benefices for the DT when the developed approach focused on humans and understands the real necessities of operators, since the real operation needs to resolve a damaged situation quickly and the reports are not updated in the same velocity, which reveals an asynchrony between the real operation and the digital data. There is a gap in the literature on studies considering this difference between real and theoretical operations and during normal and damaged situations.

Imagine a company that operates in the Brazilian electrical system, this company has some generating units such as a wind farm, some hydroelectric, and a solar farm, but it also has some substations and transmission lines. This company requests a DT for your wind farms, you know the technical challenges that we previously showed, but who is your user? What does the operator expect? In a critical situation, what is the real process? Those questions are not simple to answer, and for this, only a complete team composed of data scientists, data engineers, developers, and designers can find better answers.

Project human-centered is not the only challenge for the Brazilian energy sector, the integration of legacy systems and a data lake not structured are other problems for the DT project, and for this, the projects normally start to create a data lake and considering integrate with the legacy systems, as the examples shown in this chapter and the articles published about the DT on Brazil [12–14]. According to the research by Wanasinghe et al. [6] about the oil and gas industry in the world, there are 11 main challenges in this area, some of which can be brought to the Brazilian energy sector, as Scope and focus [15–18]; Cyber security [19–22]; Data storage and analytics [23–26]; Business model, people, and policies [17, 20, 21, 27]; Incremental vs. disruptive [21].

Companies in the energy area have been changing their view regarding the importance of designers and the user experience in their projects, but it is still common to find projects that the client does not think this is important. Other challenges are the number of legacy systems and the decoupling of real operations and data.

6. Conclusions

Brazil is a very large country, with different types of terrain, climates, and ecosystems that bring natural challenges to the electricity infrastructure of the country. With numerous generating units, substations, and extensive transmission lines scattered across the nation, the adoption of technologies like Digital Twin, IoT, Artificial Intelligence, and Big Data becomes crucial for effective asset management. Furthermore, the government's investments and initiatives in this sector promote the modernization of systems and equipment through the implementation of laws and national plans.

Although some project-specific details were not provided in this chapter, its primary objective was to showcase the challenges faced in developing real projects when companies lack the technical maturity required for a comprehensive Digital Twin implementation. The successful deployment of these technologies involves overcoming hurdles related to technology, user-centric system usability, and data availability. In addition to the articles cited, the examples presented in this chapter serve as real-world illustrations of applications that were developed to address specific challenges in the Brazilian energy sector, such as the absence of data lakes, cyber security concerns, integration with legacy systems, and operational studies.

In conclusion, the Brazilian energy sector is witnessing substantial growth and demands increased technological support to sustain and propel this progress. Energy companies in Brazil are actively pursuing the development of Digital Twins to improve equipment maintenance, enhance energy generation control, and streamline internal processes. However, it is crucial to recognize that certain companies may need to reassess their preparedness for Digital Twin implementation. By addressing any existing gaps in knowledge, infrastructure, or organizational readiness, these companies can effectively leverage the potential of Digital Twins to drive efficiency and innovation in the energy sector.

Author details

Eldrey Seolin Galindo* and Urbano Chagas
Centro de Estudos e Sistemas Avançados do Recife (CESAR), Recife, Brazil

*Address all correspondence to: eldrey.galindo@cesar.org.br

References

[1] CIA. The World Factbook. Available from: https://www.cia.gov/the-world-factbook [Accessed: 09 June 2023]

[2] ANEEL National Agency of Electricity. Sistema de geração de Informação ANEEL SIGA (Portuguese). Available from: https://dadosabertos.aneel.gov.br/ [Accessed: 13 March 2023]

[3] CCBC – Chamber of Commerce Brazil-Canada. Brazil is living a favorable moment for modernization of the electric sector. Available from: https://l1nq.com/brazil-modernization-of-the-ele [Accessed: 16 June 2023]

[4] Brazil. Lei N 13.755, Dez. Brasilia: DF Diário Oficial da União; 2018

[5] MME. Plano nacional de energia 2030. Ministério de Minas e Energia; colaboração Empresa de Pesquisa Energética. Brasília: MME; EPE; 2007

[6] Wanasinghe TR et al. DT for the oil and gas industry: Overview, research trends, opportunities, and challenges. IEEE Access. 2020;**8**:104175-104197. DOI: 10.1109/ACCESS.2020.2998723

[7] International Electrotechnical Commission. Available from: https://iec.ch/homepage [Accessed: 27 April 2023]

[8] ONS. Sistema Interligado Nacional – Rede de Operação (Portuguese). Available from: http://sindat.ons.org.br/SINDAT/Home/ControleSistema [Accessed: 02 June 2023]

[9] Inductive Automation: What is SCADA? Available from: https://inductiveautomation.com/resources/article/what-is-scada [Accessed: 02 June 2023]

[10] José Da Silva M, Melo De Souza S, Cavalcante De Lucena I, Da H, Santiago C, Galindo ES, et al. Data mining applied to abnormality prediction in electrical substation transformers. 2021. DOI: 10.37118/ijdr.23097.10.2021

[11] Pinto SCD, Villeneuve E, Masson D, Boy G, Baron T, Urfels L. DT design requirements in downgraded situations management. IFAC-PapersOnLine. 2021;**54**(1):869-873. DOI: 10.1016/j.ifacol.2021.08.102

[12] Fernandes SV, João DV, Cardoso BB, Martins MAI, Carvalho EG. DT concept developing on an electrical distribution system—An application case. Energies. 2022;**15**:2836. DOI: 10.3390/en15082836

[13] Hayashi VT, Arakaki R, Fujii TY, Khalil KA, Hayashi FH. B2B B2C architecture for smart meters using IoT and machine learning: A Brazilian case study. In: IEEE - 2020 International Conference on Smart Grids and Energy Systems (SGES), Perth, Australia. 2020. pp. 826-831. DOI: 10.1109/SGES51519.2020.00152

[14] Tjønn F. Are "digital twin through the life of a field." In: Paper Presented at the Abu Dhabi International Petroleum Exhibition & Conference, Abu Dhabi, UAE. November 2018. DOI: 10.2118/193203-MS

[15] Dattilo A, Vanderburg A, Shallue CJ, Mayo AW, Berlind P, Bieryla A, et al. Identifying exoplanets with deep learning. II. Two new super-earths uncovered by a neural network in K2 data. The Astronomical Journal. 2019;**157**(5):169. DOI: 10.3847/1538-3881/ab0e12

[16] Morais D, Waldie M. How to implement tech in shipbuilding: Charting the course to success. In: SNAME Maritime Convention.

Providence, RI, USA: The Society of Naval Architects and Marine Engineers; 2018

[17] Parrott A, Warshaw L. Manufacturing meets its match. In: Deloitte Development – Industry 4.0 and the Digital Twin, New York, NY, USA, Tech. Rep. 2017. Available from: https://www2.deloitte.com/us/en/insights/focus/industry-4-0/digital-twin-technology-smart-factory.html

[18] Parrott A, Warshaw L. Industry 4.0 and the DT. New York, NY, USA: Deloitte Development LLC; 2017

[19] Zornio P. The control room is anywhere and everywhere: Putting the industrial internet of things to work offshore and beyond. In: Paper Presented at the Offshore Technology Conference, Houston, Texas, USA. April 2018. DOI: 10.4043/28943-MS

[20] Sylthe O, Thornton B. The impact of digitalization on offshore operations. In: Paper Presented at the Offshore Technology Conference, Houston, Texas, USA. April 2018. DOI: 10.4043/28689-MS

[21] Das R, Morris T. Modeling a midstream oil terminal for cyber security risk evaluation. IFIP Advances in Information and Communication Technology. 2018;**542**:149-175. DOI: 10.1007/978-3-030-04537-1 9

[22] Mittal A, Slaughter A, Zonneveld P. Protecting the connected barrels—Cybersecurity for upstream oil and gas. New York, NY, USA: Deloitte Development LLC; 2017. Tech. Rep. Available from: https://www2.deloitte.com/tr/en/pages/energy-and-resources/articles/oil-and-gas-cybersecurity.html

[23] Fujii TY, Hayashi VT, Arakaki R, Ruggiero WV, Bulla R, Hayashi FH, et al. A DT architecture model Applied with MLOps techniques to improve short-term energy consumption prediction. Machines. 2022;**10**:23. DOI: 10.3390/machines10010023

[24] Sharma P, Hamedifar H, Brown A, and Green R. The dawn of the new age of the industrial Internet and how it can radically transform the offshore oil and gas industry. In: Proc. Offshore Technol Conferences. 2017. pp. 1-7

[25] Daily J, Peterson J. Predictive maintenance: How big data analysis can improve maintenance. In: Supply Chain Integration Challenges in Commercial Aerospace. Cham, Switzerland: Springer; 2017. pp. 267-278

[26] Ottonelli J, Cruz U, Rosa A, Andrade JC. Oportunidades e desafios do setor de energia solar fotovoltaica no Brasil. Revista Econômica Do Nordeste. version. 2021;**52**:8-26

[27] Ottonelli J, Cruz U, Rosa A, Andrade J. Opportunities and challenges of the photovoltaic solar energy in Brazil. Review Economic NE, Fortaleza. 2021; **52**(4):8-26